RAND COUNTERINSURGENCY STUDY • PAPER 4

OCCASIONAL PAPER

Money in the Bank

Lessons Learned from Past Counterinsurgency (COIN) Operations

Angel Rabasa, Lesley Anne Warner, Peter Chalk, Ivan Khilko, Paraag Shukla

Prepared for the Office of the Secretary of Defense

NATIONAL DEFENSE RESEARCH INSTITUTE

The research described in this report was prepared for the Office of the Secretary of Defense (OSD). The research was conducted in the RAND National Defense Research Institute, a federally funded research and development center sponsored by the OSD, the Joint Staff, the Unified Combatant Commands, the Department of the Navy, the Marine Corps, the defense agencies, and the defense Intelligence Community under Contract W74V8H-06-C-0002.

Library of Congress Cataloging-in-Publication Data is available for this publication.

ISBN 978-0-8330-4159-3

The RAND Corporation is a nonprofit research organization providing objective analysis and effective solutions that address the challenges facing the public and private sectors around the world. RAND's publications do not necessarily reflect the opinions of its research clients and sponsors.

Published 2007 by the RAND Corporation
1776 Main Street, P.O. Box 2138, Santa Monica, CA 90407-2138
1200 South Hayes Street, Arlington, VA 22202-5050
4570 Fifth Avenue, Suite 600, Pittsburgh, PA 15213-2665
RAND URL: http://www.rand.org/
To order RAND documents or to obtain additional information, contact
Distribution Services: Telephone: (310) 451-7002;
Fax: (310) 451-6915; Email: order@rand.org

Preface

This paper is a product of one of several RAND Corporation research projects examining U.S. political and military capabilities for fighting a spectrum of current and future insurgency threats. It should be of interest to academics, policymakers, military science specialists, intelligence analysts, and laypersons within the United States and elsewhere who may be interested in learning the applicability of lessons from past counterinsurgency (COIN) operations to the insurgencies the United States faces today and may face in the future. The six cases profiled in this paper include the Philippines (1899–1902), Algeria (1954–1962), Vietnam (1959–1972), El Salvador (1980–1992), Jammu and Kashmir (1947–present), and Colombia (1963–present). They were selected to explore COIN operations in regions with varied characteristics relating to geography, historical era, outcome, type of insurgency, and the level of U.S. or foreign involvement, among others. The issues addressed in this paper pertain to the success or failure of several counterinsurgency operations, the counterinsurgents' ability to innovate and adapt, and the need for a way to recognize the threat and determine what is needed to confront it. The authors hope that this paper will add to the ever-growing body of literature on COIN and will abet the development of tactics, techniques, and procedures (TTPs) for COIN in addition to those cited in the newly released U.S. Army and U.S. Marine Corps Counterinsurgency Field Manual (FM 3-24/ MCWP 3-33.5).

This paper is not intended to be a comprehensive analysis of insurgency. Instead, it examines a handful of insurgencies to determine which TTPs the insurgents and counterinsurgents employed and then develops some general conclusions on counterinsurgency operations that could be applied to the 21st century. To this end, the authors have included theories and lessons learned from the past, as well as contemporary debates on the topic, which may not necessarily pertain to the lessons learned from these particular case studies.

Although this paper had several authors, Lesley Anne Warner, as the lead author, was responsible for collating the various chapters, ensuring cohesiveness and continuity in the case study analyses, and enumerating overall lessons from the various counterinsurgency operations that are described here.

This research was sponsored by the U.S. Department of Defense and was conducted within the International Security and Defense Policy Center of the RAND National Defense Research Institute, a federally funded research and development center sponsored by the Office of the Secretary of Defense, the Joint Staff, the Unified Combatant Commands, the

Department of the Navy, the Marine Corps, the defense agencies, and the defense Intelligence Community.

For more information on RAND's International Security and Defense Policy Center, contact the Director, James Dobbins. He can be reached by email at James_Dobbins@rand.org; by phone at 703-413-1100, extension 5134; or by mail at the RAND Corporation, 1200 South Hayes Street, Arlington, VA 22202-5050. More information about RAND is available at http://www.rand.org/.

Contents

Figures and Tables

Figures

Tables

Summary

The Global War on Terror (GWOT) is being waged in multiple theaters possessing a wide spectrum of social dynamics, regional relationships, histories, political cultures, strengths and weaknesses, and salient grievances. As insurgent threats evolve and assume new forms, the United States must also evolve in its ability to counter potentially prolonged threats in several parts of the world. Because of the potential for global reach in contemporary insurgencies, the ability to draw on lessons learned from past counterinsurgency (COIN) operations using different historical cases can be valuable, helping current and future leaders prevent a repetition of mistakes and elucidating a foundation on which to build contemporary responses. Despite the need to look to the past for clues on how to proceed at present or in the future, it is also important not to generalize, making lessons learned not a loose analogy but a perfectly matching antidote. Rather than disregarding successes and failures as phenomena of the past or attempting to shove round lessons into square counterinsurgencies, strategists must consider a range of possible responses.

This paper analyzes six COIN case studies from the 19th and 20th centuries in which insurgent and counterinsurgent strengths and weaknesses are examined for their contributions to the outcomes of the conflicts, if they have been resolved as of this writing. The cases profiled in this paper are the Philippines (1899–1902), Algeria (1954–1962), Vietnam (1959–1972), El Salvador (1980–1992), Jammu and Kashmir (1947–present), and Colombia (1963–present). These cases were selected because of the potentially valuable lessons that can be drawn from them for future COIN operations and because they demonstrate the application of some of the methods detailed in the U.S. Army and U.S. Marine Corps Counterinsurgency Field Manual (FM 3-24/ MCWP 3-33.5) released in December 2006. As the reader will find, in addition to the various tactics, techniques, and procedures (TTPs) used to combat these insurgencies, these cases exhibit variations and commonalities in such characteristics as outcome, historical era, geographic spread, type and organization of the insurgency, and the level of foreign intervention, among others.

The Philippines (1899–1902)

The insurgency in the Philippines did not have a strong base of support among the population, because some Filipinos wanted provincial autonomy, whereas the insurgents' goal was centralized government. The insurgency was highly factionalized with competing goals and it often alienated potential supporters in the population by levying taxes on them and using violence against those discovered to be cooperating with the Americans. The insurgents were also weakened by the fact that they were ill-trained, transitioned to guerrilla tactics too late in the war to have a significant effect, and were unable to obtain sanctuary in nearby countries or to arrange for any possible influx of supplies and manpower as a result of the country's island geography.

Although the insurgents outnumbered the Americans and were often able to disappear into the population, the well-trained U.S. soldiers were able to defeat the insurgency despite their own misconceptions about the conflict, their unfamiliarity with the terrain, and the brutal tactics they employed to put down the insurrection. Many U.S. soldiers had learned how to fight a war of this type from their experiences in Puerto Rico and Cuba, as well as during the wars against the American Indians between the 1860s and the 1890s. From these conflicts, they learned to separate the population from the insurgents and to ration food to decrease the population's incentive to share food with the insurgents. The counterinsurgents also participated in contingency operations during which they would help create and maintain infrastructure. The Americans also made it inescapably clear that collaboration with the insurgents would be severely punished. To help restore law and order to the archipelago, the United States created armed local indigenous forces who were instrumental in capturing the insurgent leader, gathering intelligence, and protecting the population from insurgent retribution.

Algeria (1954–1962)

In Algeria, the insurgent goal was to establish an independent state within the framework of the principles of Islam, although most of the population remained ambivalent until the Front de Libération Nationale (FLN) initiated a campaign of discrete urban terrorism. The beginning of this campaign instigated a French overreaction targeting the Algerian population as a whole with such brutality that the FLN's cause immediately gained popularity. The FLN's targeting of civilian-centric venues in Algiers' European sector resulted in the French employing extrajudicial means to detain, interrogate, and torture suspected insurgents. The draconian measures the French took to quell the insurgency eventually drove even unaffiliated moderates into the outstretched arms of the FLN. Once news of the institutionalized regime of torture was made known abroad, French public support for the war plummeted.

Eventually, the French realized that they needed to gain the support of the population through humanitarian assistance and secure Algeria's borders to eliminate the influx of external support to the insurgents. Ultimately, they sought to persuade the population that they fared better under French rule than as an independent nation. Although the second half of France's COIN strategy was successful, it was compromised by the degree to which France had attempted to pacify the country through brute force. With the loss of public support for the war at home, France was forced to grant Algeria independence after winning the military war but losing the political one.

Vietnam (1959–1972)

The part of the insurgency in Vietnam covered in this paper was a continuation of the Vietnamese War for independence from the French (1945–1954), from which a communist North Vietnam and a U.S.-backed South Vietnam emerged. North and South Vietnamese communists began their infiltration and indoctrination of cadres in the south and accelerated these efforts in the aftermath of the coup against Ngo Dinh Diem. Throughout the war, the insurgents emphasized the war's political nature, using established networks to gain the support of the population and creating mass associations as a vehicle for political indoctrination. Acting as a shadow government, they were able to provide social services, enact land reform, and make those in the population feel that they had a stake in supporting them. Knowing that the South Vietnamese government was too weak to protect its population, insurgents also used discriminant terrorism to maintain control of their own cadres and the population at large. By the time U.S. combat troops arrived, the situation in South Vietnam had deteriorated to the point that U.S. involvement could not be restricted to counterinsurgency alone.

By the mid-1960s, pacification programs carried out under Diem were placed on the back burner, as U.S. and South Vietnamese security forces struggled to regain control of the military situation. The inability of indigenous security forces to shoulder an adequate amount of combat responsibility, inefficiencies in gathering intelligence from the population to target the insurgent infrastructure, and the unwillingness of the South Vietnamese government to build a political base perpetuated the spiral of chaos leading up to the Tet Offensive in 1968. Because of the severity of enemy losses during Tet, the counterinsurgents were able to intensify their pacification efforts and achieve moderate success. It was during this period that the United States was able to achieve unity of command with Civil Operations and Revolutionary Development Support (CORDS) and that the South Vietnamese government, in an effort to build a political base outside Saigon, enacted a program of land reform. Despite these innovations and reforms, the south was overrun by a conventional invasion from North Vietnam in 1975. Thus, the true periods of COIN and pacification in Vietnam occurred between 1959–1963 and 1968–1972.

El Salvador (1980–1992)

The insurgency in El Salvador emerged from political-criminal activities, such as kidnappings and assassinations, and eventually evolved into guerrilla warfare. Although there was no mass support for such an insurgency, the insurgents were able to field a large number of fighters relative to the Salvadoran security forces, with members spanning the political spectrum. Although unified into one insurgent group, there was substantial disagreement as to the doctrine and identity of the movement, which severely compromised its strength. Because of the country's rugged terrain and unregulated border with Honduras, the insurgents were able to enjoy sanctuary, as well as a steady flow of support from Cuba and Nicaragua, until the fall of the Soviet Union.

As a result of a series of free elections, the Salvadoran government has been awarded broad popular support and, thus, political legitimacy. To build on its legitimacy, the government implemented civic action programs to rebuild social and economic infrastructures and free the army to pursue insurgents. Additionally, a train and equip program run by the United States helped retrain the Salvadoran Army to fight the insurgency, although direct U.S. involvement was kept to a minimum. The government's lack of control over death squad activity eroded domestic and international support, and uncertainty over continued U.S. support resulted in less-effective warfighting. The insurgency ended with a negotiated compromise in which the insurgents were given a stake in the political future of the country.

Jammu and Kashmir (1947–Present)

The insurgency in Jammu and Kashmir (J&K) has been ongoing for over half a century and has been sustained by support from Pakistan and by an influx of foreign fighters who may have links to al Qaeda. The various competing factions draw members from the ranks of other insurgent organizations and their cause is to establish a fundamentalist theocracy. The insurgents are mainly rural, because there are few security forces in those areas, and they do not provide social services or any form of informal government to local civilians. They frequently employ terrorism indiscriminately to force loyalty and instill fear in the population.

The Indian government, learning from British lessons during the Malayan Emergency as well as from its own experience with the Liberation Tigers of Tamil Eelam (LTTE), has been rather successful at militarily managing the insurgency. The government has created specially trained units to execute COIN, separated the civilians from the insurgents, protected the population, and restricted the use of airpower and firepower to reduce civilian casualties. It has also engaged in civic action to ensure amicable relations with the population and to encourage cooperation in gathering intelligence. The insurgency is ongoing largely because the insurgents enjoy sanctuary in Pakistan and a political solution has not yet been developed and applied.

Colombia (1963–Present)

The Colombian insurgent groups emerged from an atmosphere of revolutionary change in which they sought to take political power by force. Over time, their income has come from kidnapping, extortion, and the local drug trade, through which they have interacted with Latin American organized crime networks to ensure a steady supply of arms. Through the movement's involvement in the drug trade, the insurgency has lost ideological cohesion, as many leaders have become more interested in personal enrichment than in the organization's political and military agenda. Furthermore, many potential domestic and international supporters have been repelled by the insurgents' involvement in the drug trade and their use of indiscriminate terrorism, and they are consequently extremely unpopular. The insurgents have also failed to challenge major population centers or sabotage vital economic assets and they have not used their vast source of income to acquire sophisticated weapons to neutralize the government's air superiority.

The Colombian government has had the advantage of political legitimacy, with a long record of freely and fairly elected civilian leadership. The government receives substantial aid from the United States and the European Union, with which they have increased the strength of their security forces, armed and trained local self-defense units, and implemented a seize-and-hold strategy to flush the insurgents from certain territories. However, the government does not have the numbers to secure the borders and maintain the seize-and-hold strategy. The insurgents enjoy sanctuary in Venezuela, Panama, and Ecuador and there is also evidence that they receive some level of support, tacit or overt, from Venezuelan president Hugo Chavez.

Some of the characteristics of the insurgencies covered in this paper can be found in Table S.1.

Conclusions

When presented with a variety of possible insurgencies, counterinsurgents may be more adept at managing the problem if they have "money in the bank"—in other words, if they can benefit from lessons learned during past COIN operations. For the sake of continuity and adaptability in the multifront Global War on Terror, counterinsurgents should approach lessons learned across past COIN operations as loose analogies. At the same time, those charged with executing COIN should avoid making generalizations that tend to form a model for COIN. Overall, seeing how counterinsurgents confronted the complexities of the insurgencies they faced in the past may enable current counterinsurgents to be more proficient at fighting a wide variety of modern insurgencies that have global reach. In the past cases of the Philippines and Vietnam and in the ongoing cases of Jammu and Kashmir and Colombia, the counterinsurgents were open to using knowledge gained from past counterinsurgency operations, which they then used to formulate TTPs for their ongoing operations. Doing so often required that they be objective critics in the face of failure and adjust their strategy accordingly.

It is important that counterinsurgents understand local dynamics so that all theaters of the conflict can be understood in context. This knowledge can help exploit cleavages and encourage competition among insurgent factions, which was done in the Philippines and, with less success, in Vietnam. In Vietnam, El Salvador, and Colombia, counterinsurgents used indigenous intermediaries with established social networks to earn the trust of the population and psychologically unhinge the insurgents. In some of these cases, the indigenous intermediaries took the form of armed civilian self-defense militias who protected their own villages from insurgent attacks. In the case of the Philippines, the creation of a well-trained and uncorrupt police force was integral to the capture of the key insurgent leader and in demonstrating that locals were being trusted to provide for and control their own security. Police are also integral to counterinsurgency operations because they are responsible for detaining and interrogating suspected insurgents, from whom they can acquire intelligence to attack the insurgent infrastructure.

Table S.1
Characteristics of Selected COIN Case Studies

Characteristic	Philippines	Algeria	Vietnam	El Salvador	Jammu and Kashmir	Colombia
Insurgent goal	Independence	Independence	Marxism	Marxism	Islamist control	Marxism
Insurgent approach	Military	Political/military	Political/military	Political/military	Military	Military
Organizational structure	Hierarchical	Medium	Hierarchical	Hierarchical	Hierarchical	Hierarchical
Sanctuary	No	Yes	Yes	Yes	Yes	Yes
Sanctuary denied	N/A	Yes	No	No	No	No
Level of foreign counterinsurgent intervention	Direct military	Direct military	Direct military	Train and equip	No	No
Foreign support	No	Yes	Yes	Yes	Yes	Yes
Counterinsurgent-to-insurgent ratio	1.3	10	2.9	4	50.0	15.3
Population-to-COIN force ratio	59.5	26.2	11.6	79.2	18.5	143.4
Outcome	COIN win	COIN loss	COIN loss	COIN win	Ongoing	Ongoing

SOURCE: Data collected by Martin Libicki based on coding by RAND researchers.

Depending on the situation, a hands-off approach is sometimes necessary to allow the host nation to learn which methods are most effective in dealing with an insurgency, considering its own strengths and limitations. With this in mind, foreign counterinsurgents should determine how best to assist the host nation in its efforts to reform, if this is necessary, to better fight the insurgency. Diversifying sources of data on the host nation and gathering information on its intelligence collection and dissemination abilities may support this effort.

As in the cases of El Salvador and Colombia, strong, competent, democratically elected leadership at all levels of government is especially helpful in situations in which both the insurgents and counterinsurgents are attempting to persuade the population not only that their form of government is legitimate but also that they will have the opportunity to improve their quality of life and the political means to express their desire for this. Efficient host nation provision of social services and employment opportunities can also demonstrate legitimacy and competence in the eyes of the population. Foreign or even host nation counterinsurgents who are not from the local area of operations should assume that they will have limited opportunities to convey their good intentions. Consequently, they may be viewed more favorably from the outset if they are perceived as contributing to progress and not to chaos. In the three cases with large foreign counterinsurgent contingents (the Philippines, Algeria, and Vietnam), as well as in Jammu and Kashmir, the counterinsurgents engaged in humanitarian actions designed to improve the lives of the population, although in some cases these actions were taken either too late or on such a small scale that they had minimal effect.

Counterinsurgents should strive for "unity of command," akin to the bureaucratic structure of the CORDS program in Vietnam, so that there is fusion and continuity among counterinsurgency programs. To facilitate this structure, bureaucracies should encourage a culture of cooperation, both in the host nation and among the foreign counterinsurgents, and have either a foreign adviser in the background or a domestic political leader to bridge this gap. In the area of operations, local autonomy for counterinsurgents may enable innovation and adaptability.

In the case of Algeria, the French were extremely adept at securing the country's borders to deny insurgents sanctuary, to minimize the influx and influence of unwanted external actors, and to sap the strength of the insurgent infrastructure. However, counterinsurgents failed in this effort in Vietnam and El Salvador, as well as in the ongoing cases of Jammu and Kashmir and Colombia. This failure has allowed insurgents to maintain the strategic initiative and recuperate mentally and physically in their sanctuaries when they feel threatened by the counterinsurgents.

Finally, counterinsurgents should analyze solutions in terms of long-term effectiveness, not short-term necessity. As demonstrated by the time spans of all the counterinsurgency operations discussed in this paper, insurgency can be a prolonged affair. In the face of long-term necessity, short-term effectiveness is often a poor substitute, especially when actions taken in the short term to solve immediate problems counteract the long-term goals of the counterinsurgency operation.

Acknowledgments

The authors wish to express their sincere thanks to the many people who sponsored, supported, and critiqued this research. This project was made possible through the support of our sponsors in the Office of the Secretary of Defense: Benjamin P. Riley III, Director, Rapid Reaction Technology Office, Chairman, Combating Terrorism Technology Task Force; and Richard Higgins, Program Manager, Technical Support Working Group. The authors would also like to thank John Gordon and William Rosenau for their comments on multiple drafts, Martin Libicki for providing useful data on characteristics of the insurgencies described in this paper, and Brian Nichiporuk and Robert Everson for performing thorough and insightful formal reviews.

Abbreviations

ANC	African National Congress
AOR	area of responsibility
ARVN	Army of the Republic of Vietnam
AUC	United Self-Defense Forces of Colombia [Autodefensas Unidas de Colombia]
BSF	Border Security Force
CAP	Combined Action Program
CDHES	Salvadoran Commission of Human Rights [Comision de Derechos Humanos de El Salvador]
CIA	Central Intelligence Agency
COIN	counterinsurgency
CORDS	Civil Operations and Revolutionary Development Support
COSVN	Central Office for South Vietnam
DRU	Unified Revolutionary Directorate [Direccion Revolucionario Unificada]
ELN	National Liberation Army [Ejército de Liberación Nacional]
ELP	Popular Liberation Army [Ejército de Liberación Popular]
ERG	Guevarista Revolutionary Army [Ejército Revolucionairo Guevarista]
ERP	People's Revolutionary Army [Ejército Revolucionairo del Pueblo]
EU	European Union
FAPU	Unified Popular Action Front [Frente de Acción Popular Unificada]
FARC	Revolutionary Armed Forces of Colombia [Fuerzas Armadas Revolucionarias de Colombia]
FARN	Armed Forces of National Resistance [Fuerzas Armadas de Resistencia Nacional]
FDR	Democratic Revolutionary Front [Frente Democrático Revolucionario]
FES	Fuerzas Especiales Selectas

FLN	Front de Libéracton Nationale
FMLN	Farabundo Martí National Liberation Front [Frente Farabundo Martí para la Liberación Nacional]
FPL	Popular Liberation Forces [Fuerzas Populares de Liberacion]
FY	Fiscal Year
GO	general order
GPRA	Gouvernement Provisoire de la République Algérienne
GVN	Government of South Vietnam
GWOT	Global War on Terror
HuM	*Harkat-ul-Mujahideen*
HUMINT	Human Intelligence
IED	improvised explosive device
IO	information operations
ISI	Inter-Services Intelligence
JeM	*Jaish-e-Mohammad*
JKLF	Jammu and Kashmir Liberation Front
LeT	*Lashkar-e-Toiba*
LoC	line of control
LP-28	Popular Leagues February 28
LTTE	Liberation Tigers of Tamil Eelam
MACV	Military Assistance Command Vietnam
MIR	Movement of the Revolutionary Left [Movimiento de Izquierda Revolucionaria]
NGO	nongovernmental organization
NLF	National Liberation Front
NATO	North Atlantic Treaty Organization
OAS	Organisation de'l Armée Secrète
OPATT	Operational Planning and Assistance Training Team
OSS	Office of Strategic Services
PCES	Communist Part of El Salvador [Partido Comunista de El Salvador]
PCN	National Conciliation Party [Partido de Conciliación Nacional]
PLO	Palestine Liberation Organization
PRI	Revolutionary Institutional Party [Partido Revolucionario Institucional]

PRP	People's Revolutionary Party
PRTC	Revolutionary Party of Central American Workers [Partido Revolucionario de los Trabajadores Centromericanos]
PSYOPs	psychological operations
RMTC	Regional Military Training Center
RR	Rashtriya Rifles
TTPs	tactics, techniques, and procedures
UN	United Nations
UNCIP	United Nations Commission for India and Pakistan
UH	Unified Headquarters
VCI	Viet Cong Infrastructure

CHAPTER ONE

Introduction

The Global War on Terror (GWOT) is being waged in multiple theaters possessing a wide spectrum of social dynamics, regional relationships, histories, political cultures, strengths and weaknesses, and salient grievances. In the post-9/11 world, many policymakers refer to GWOT as the "Long War," in which proficiency in counterinsurgency (COIN) operations could be the difference between defeat and victory on a timetable that is more in harmony with U.S. capabilities to counter prolonged threats in several parts of the world. As global threats are in a state of constant flux, U.S. capabilities must strive to stay ahead of the curve. Because of the potential for global reach in contemporary insurgencies, the ability to draw on lessons learned can be a valuable resource, not only in comprehending how certain methods have been applied in various settings but also in discovering possible strands of continuity during what Prussian military theorist Carl von Clausewitz (1968) described as the "fog of war." Seeing general trends and outcomes across cases from the past should help prevent repetition of mistakes and elucidate a foundation on which to build contemporary responses.

Despite the need to look to the past for suggestions on how to proceed, perhaps the greatest lesson that can be learned from looking at past COIN campaigns for insights on current and future campaigns is not to generalize. Insurgencies and counterinsurgencies are not clones; the solutions and problems may or may not be transferable between cases. Seeing lessons learned not as a loose analogy but as a perfectly matching antidote can be a rather costly oversight. In the search for lessons learned and unlearned, counterinsurgent strategists should regard the approach to these lessons as more consistent with the qualities of flypaper than Teflon, although neither extreme is ideal. Rather than disregarding successes and failures as a phenomenon of the past or attempting to shove round lessons into square counterinsurgencies, strategists must be open to multiple possibilities.

This paper is one of several RAND Corporation research products of a large project tasked with determining future political and military capabilities for fighting a spectrum of insurgencies. While taking a broad look at the phenomenon of insurgency, the authors decided to zoom in and take a more intimate look at which tactics brought insurgencies and counterinsurgencies success and failure. In undertaking this study, the authors hoped to derive important insights from a collage of insurgencies—big and small, ongoing and completed, insurgent and counterinsurgent victory, and with or without U.S. involvement.

For the purpose of this paper, the authors used the Central Intelligence Agency (CIA) definition of insurgency, which states that

> **Insurgency** is a protracted political-military activity directed toward completely or partially controlling the resources of a country through the use of irregular military forces and illegal political organizations. Insurgent activity—including guerrilla warfare, terrorism, and political mobilization, for example, propaganda, recruitment, front and covert party organization, and international activity—is designed to weaken government control and legitimacy while increasing insurgent control and legitimacy. The common denominator of most insurgent groups is their desire to control a particular area. This objective differentiates insurgent groups from purely terrorist organizations, whose objectives do not include the creation of an alternative government capable of controlling a given area or country.[1]

For the purpose of this paper, the authors also used the definition of counterinsurgency from the *Department of Defense Dictionary of Military and Associated Terms*, which states that

> **Counterinsurgency** consists of those military, paramilitary, political, economic, psychological, and civic actions taken by a government to defeat insurgency.[2]

The cases analyzed in this paper are

- The Philippines (1899–1902) by Ivan Khilko
- Algeria (1954–1962) by Peter Chalk
- Vietnam (1959–1972) by Lesley Anne Warner
- El Salvador (1980–1992) by Angel Rabasa
- Jammu and Kashmir (1947–present) by Paraag Shukla
- Colombia (1963–present) by Angel Rabasa.

These cases were selected for multiple factors, among them, geographical distribution, outcome, U.S. and foreign involvement, and historical era. The Philippines case was selected because it was one of the few clear-cut U.S. COIN successes, although it can be perceived as less relevant to 21st-century insurgencies because of the changes in technology over the past century. However, like the U.S.-led invasion of Iraq in 2003, the Philippines case is an example of an invasion by foreigners who then faced an unexpected uprising and were forced to adapt to it. Their methods of population control and system of rewards and punishment ("carrot and stick") are examples of pacification that can be used as an alternative to the other well-known approaches of "winning hearts and minds" or "transformation."[3]

The authors selected Algeria as a case study because it was a prime example of a COIN operation in which the counterinsurgents were able to prevail militarily but subsequently lost the political war, especially on the home front, as a result of the brutality used to put down

---

[1] Central Intelligence Agency (n.d.).

[2] Joint Chiefs of Staff/Department of Defense (2007).

[3] "Winning hearts and minds" refers to counterinsurgents engaging in civic action to gain the cooperation of the population through their good deeds and to discourage cooperation with the insurgents. "Carrot and stick" is a strategy of incentives and punishments intended to reward or punish the population for its cooperation or lack thereof. "Transformation" is a long-term strategy to reform the social and governmental norms of a country so that grievances can be addressed legally, as opposed to the population having to resort to insurgency (Gompert and Gordon, unpublished research).

the insurgency. Algeria is also a classic example of how insurgent tactics prompted counterinsurgents to react with such overwhelming force that they alienated the previously unaffiliated members of the population and drove them into the arms of the insurgency.

The insurgency in Vietnam was chosen not only because it was a watershed in U.S. history but also because it was a definitive moment in the evolution of military culture with regard to fighting political-military wars. This case is also a prime example of the effectiveness of a joint political and military approach to COIN through the Civil Operations and Revolutionary Development Support (CORDS) program, although it might have proven more effective had there been an early and accurate assessment of the situation on the ground. Although the counterinsurgents lost in this case, Vietnam is a good example of how pacification was ultimately so successful that North Vietnam was forced to launch a conventional invasion of South Vietnam in order to win the war.

The El Salvador case was selected because it was an example of a rather successful COIN operation and a post-Vietnam experiment in which the United States was able to limit its role to train and equip. Because of the favorable outcome of this conflict, this case is also used as an example of how the United States, playing a relatively indirect role, was able to train the Salvadoran Army to fight the insurgency more efficiently, which was instrumental in breaking its momentum. The host nation's willingness to participate in a series of elections added to its legitimacy and showed that it was committed to democratic governance, which may have contributed to the government's eventual victory.

The authors chose Jammu and Kashmir to include in this paper because of its Islamist element, the long duration of its insurgency, and the role of external support in sustaining the insurgency. In addition, Jammu and Kashmir is an example of an insurgency in which the insurgents have been more concerned with securing their own interests and less concerned with creating a social network for political activism, unlike many of the other insurgencies discussed in this paper.

Within the scope of the insurgencies analyzed here, Colombia represents another example of an insurgency that has not been centered on gaining the support of the population but rather began with ideological roots that were soon fragmented by the very activities that sustained it. The authors chose this case because of its long duration, the insurgency's transformation over this span of time, and the government's relative success in enacting measures to keep the insurgency at bay.

Some of the characteristics of these insurgencies can be found in Table 1.1.

In each case study, insurgent and counterinsurgent tactics and strategies have been analyzed, as well as the strengths and weaknesses of each side and how they contributed to the outcome of the conflict. In conclusion, the authors sought to acknowledge the lessons learned from each conflict and the similarities and differences among methods that were successful in combating a spectrum of insurgencies. In addition, the authors sought to articulate general conclusions on COIN operations, whether addressed in these particular case studies or not. Although most of the insurgencies discussed in this paper are part of the classical genre of insurgency, it is still important to understand the complexities of these cases especially as the United States prepares to wage GWOT for the long haul. The ability to draw on these

Table 1.1
Characteristics of Selected COIN Case Studies

Characteristic	Philippines	Algeria	Vietnam	El Salvador	Jammu and Kashmir	Colombia
Insurgent goal	Independence	Independence	Marxism	Marxism	Islamist control	Marxism
Insurgent approach	Military	Political/military	Political/military	Political/military	Military	Military
Organizational structure	Hierarchical	Medium	Hierarchical	Hierarchical	Hierarchical	Hierarchical
Sanctuary	No	Yes	Yes	Yes	Yes	Yes
Sanctuary denied	N/A	Yes	No	No	No	No
Level of foreign counterinsurgent intervention	Direct military	Direct military	Direct military	Train and equip	No	No
Foreign support	No	Yes	Yes	Yes	Yes	Yes
Counterinsurgent-to-insurgent ratio	1.3	10	2.9	4	50.0	15.3
Population-to-COIN force ratio	59.5	26.2	11.6	79.2	18.5	143.4
Outcome	COIN win	COIN loss	COIN loss	COIN win	Ongoing	Ongoing

SOURCE: Data collected by Martin Libicki based on coding by RAND researchers.

lessons, as insurgents draw on lessons learned from past insurgencies, may well be the factor determining which side can most quickly outmaneuver the other.

CHAPTER TWO

The Philippines (1899–1902)

Ivan Khilko

Origins and Characteristics of the Insurgency

The Filipino insurgency has its roots in the waning days of Spanish occupation as a revolutionary organization called the Katipunan,[1] started in 1892. The Katipunan accomplished little during their time under the Spanish, largely because there was a great deal of infighting within the organization. This is best exemplified by the assassination of the organization's leader, Andres Bonifacio, in 1897 by Emilio Aguinaldo, a member of the politically active *principale* class of landed gentry. By removing his rival, Aguinaldo assumed power of the largest rebel factions in the country and would go on to lead the insurgency against the United States.

For the most part, Aguinaldo was able to centralize the insurgency on Luzon, the largest island in the archipelago, ordering his followers to go into the mountains; in November 1897, he issued a decree that they would adopt guerrilla tactics against the Spanish. In response to this, the Spanish concluded the Pact of Biak-na-Bato with him a month later, which ostensibly called for reparations for injured Filipinos and reforms beneficial to the indigenous population. (See Figure 2.1 for a map of the Philippines.)

After being given 400,000 pesos by the colonial government, Aguinaldo was sent off to Hong Kong with a group of his followers, allowing the Spanish to only nominally implement the reforms discussed. The departure also did not end the Filipino revolutionary efforts and only bolstered Aguinaldo's myth throughout the archipelago.[2] Admiral George Dewey of the U.S. Navy brought Aguinaldo back from Hong Kong in May 1898, thinking that he would be a good ally against the Spanish for the impending U.S. takeover of the Philippines. At this point, the relationship between Aguinaldo and the United States begins to blur, because President William McKinley gave unclear instructions to Dewey as to the future purpose of the U.S. presence on the archipelago.

A number of disagreements at key junctures, such as deciding who had responsibility for capturing Manila from the Spanish, turned Aguinaldo against the United States. He called a constitutional convention in January 1899 that accomplished very little, as the various

[1] "A blend of revolutionary rhetoric, nationalist ideals, Tagalog ethnocentrism, and secret society rituals" (Linn, 1989).

[2] Linn (1989).

Figure 2.1
The Philippines

RAND *OP185-2.1*

classes of Filipino society presented strongly divergent interests.[3] After the start of U.S. hostilities against revolutionary forces in February 1899 in a Manila suburb, Aguinaldo realized that he had to significantly reorganize the Revolutionary Army and opted to do so by mixing conventional and guerrilla forces. Another important player on the Filipino side, Lieutenant General Antonio Luna, attempted to play up the conventional warfare side of Aguinaldo's strategy but, because of his erratic nature on and off the battlefield, he was assassinated by Aguinaldo's bodyguards in June 1899.

In December 1899, Aguinaldo decided to abandon conventional tactics and turn wholly to guerrilla warfare, realizing that the Revolutionary Army was no match for the much-better-trained U.S. forces led by Major General Elwell Otis. In Luzon, Aguinaldo divided insurgent forces into partisans and militia, with the partisans focusing on disrupting U.S. progress on the island and attacking towns that capitulated to U.S. forces and the militia fighting as part-time guerrillas responsible primarily for helping out the partisans. Members of the two often switched between groups arbitrarily and the organization also contained strata that included town government representatives and tax collectors. Because Aguinaldo had the *principales* on

[3] Linn (1989).

his side, they were able to help him win over a significant number of townsfolk to the guerrillas' side, often encouraging them to join the militia.

The year 1900 proved to be a turning point in the war, with General Arthur MacArthur replacing General Otis in May of that year as U.S. commander for the Philippines. With a broader understanding of the rebellion, MacArthur approved tactics that ensured that the United States could secure the cooperation of the Filipinos through visible threat of force against rebel collaborators. Reaching out to the bases of power of the wealthy *ilustrado* businessmen and the *principales*, MacArthur was ultimately able to win the support of both of these groups and that of the people they controlled. By the time Major General Adna Chaffee was appointed U.S. commander of the Philippines in July 1901, Aguinaldo had been captured and the rebel movement had lost most of its drive and organization. After putting down rebellions in South Luzon and on the island of Samar, the war was officially declared finished in June 1902.

Strengths of the Insurgents

1. *Able to navigate terrain well.* The rebels used their knowledge of the difficult geography of the Philippines to their advantage, often retreating to hideouts in the mountains of North Luzon or leading the U.S. Army into the jungle where fighting was extremely tough. In so doing, they were able to, for the most part, fight battles in places where a conventional army would be stuck, making it hard to defeat a small force of guerrillas with superior knowledge and capabilities in compromising terrain.
2. *Blended in with civilian population.* As in any guerrilla conflict, the greatest strength of the rebels was that there was no apparent dividing line between combatants and civilians. When fighting was over, combatants could very easily slip back into villages where they would be indistinguishable from the noncombatant population. This was extremely frustrating for U.S. Army commanders who were well aware that a large number of the people they were helping in the villages and who seemed to be siding with them were in fact likely to be spies for the rebel leadership, or enemy combatants.[4]
3. *Vastly outnumbered U.S. forces.* Most modern histories put the number of Filipino insurgents around 100,000. Although the total number of U.S. troops was nominally higher (126,468), the number of combat troops was, at any given time, closer to 40,000.[5] Thus, the Filipinos had a significant advantage in the number of fighters they could commit to any given conflict.

[4] Captain Delphey T. E. Casteel, writing to his wife in October of 1900: "One day we may be fighting with thousands of their people [and] the next day you can't find an enemy, they are all 'amigos'" (Linn, 1989).

[5] Deady (2005).

Weaknesses of the Insurgents

1. *Lack of strong leadership.* Most Filipino commanders had generally received no formal military training whatsoever under the Spanish and those who had, like Aguinaldo, generally had experience fighting off only small groups of bandits, which was not particularly useful in a large-scale conflict. This proved to be a significant problem in terms of strategizing, particularly in the early stages of the conflict when the insurgents were still pursuing conventional tactics. The decision for the wholesale shift to guerrilla tactics was likely taken too late, after the rebels had lost a significant number of men and the control of key areas of Luzon. The lack of coordination at the command level is also highlighted by the fact that quite often the Filipino commanders seemed to be driving at diverging aims, as evidenced by the assassination of General Luna in 1899. Aguinaldo could hardly hope to centralize the insurgents, as he was in hiding until September 1900, by which time the insurgency had mutated into a number of independent battalions with little communication among them.
2. *Weak support base.* Many of the tactics officially sanctioned by the rebel leadership, such as intimidation and monetary extortion from the impoverished peasantry, served to decrease potential assistance from the populace. Aguinaldo's choice of the *principales* and the Tagalogs (he was a member of both groups) as his natural support base gave him a significant lead against the U.S. Army, but this was too narrow to build a lasting following in the countryside.[6] Through concentrating mainly on large landowners and promising lucrative political incentives after victory to only a small class of rich Filipinos, Aguinaldo alienated the lower classes of Filipino society.[7] U.S. forces were able to win their support through contingency operations by showing that they were actually there to help those who supported them. Ultimately, Aguinaldo's oversight proved crucial to the U.S. campaign, as the Army was able to use the *ilustrados* to influence a significant portion of the peasant population and prevent them from siding with the rebels. The fact that Aguinaldo banked on a change of U.S. presidency also led to his undoing, as this was hardly something he could realistically hope to influence from his position in the Philippines, even considering the strong anti-imperialist sentiment in the United States at the time. Also, the aims of the insurgency after winning—attempting to establish a centralized government with power over the whole archipelago—seemed to run counter to what many Filipinos wanted—greater provincial autonomy.[8] Another possible base of support that the Revolutionary Army disregarded were the *ladrones*—roving gangs of thieves, derived from the peasant class, that terrorized the countryside. Given the *principale*-centric nature of the insurgency, Aguinaldo and his commanders saw the *ladrones* as a threat and often fought with these potential allies who were adept

---

[6] Deady (2005).

[7] "With the exception of Malvar and a few others, the insurgent jefes refused to countenance a social revolution or to advocate radical policies—such as land distribution—that might have attracted the peasantry" (Linn, 2000).

[8] Linn (2000).

at blending in with the population and evading capture as well as skillful in navigating the land.

3. *Terrorism and oppression of the local population.* The Revolutionary Army resorted to terrorism toward the middle of the conflict, as they saw that many in the population were cooperating with the Americans. These tactics grew more and more violent: Rural collaborationists were slaughtered in ever-growing numbers, creating a culture of intimidation that served to turn much of the rural population against Aguinaldo's troops. In addition to this, the fact that the terrorists imposed taxes on often-impoverished local populations and sometimes resorted to force or outright theft to replenish dwindling supplies led to animosity toward them in the countryside.
4. *Island geography as a limiting factor.* The Revolutionary Army, as is often the case with island insurgent groups, had a hard time fighting the U.S. Army precisely because of the restrictive archipelagic geography of its homeland. Each island, although large enough for several military theaters, did not offer much in the way of hiding places, forcing militants to constantly hide among the population after an attack and making it very hard to regroup and recuperate without being discovered. The isolation of the islands, coupled with the U.S. Navy blockade of the archipelago, made it difficult for the insurgents to coordinate actions[9] among themselves. Also, even if the Revolutionary Army had had a major outside supporter (which it did not), it would have been difficult to get support onto the islands, further limiting its chances for victory against the United States.
5. *Poorly trained and armed corps of soldiers.* The lack of military training in the lower ranks of the Revolutionary Army was also a significant problem, as the insurgents were no match for the formally trained U.S. troops.[10] This is indicated by the significantly higher number of casualties suffered by the Filipinos in nearly any conflict with U.S. forces. Additionally, a number of incidents were recorded throughout the war in which rebels made serious tactical mistakes, such as forfeiting large quantities of weapons in battle. As far as armaments go, the Filipino troops' rifles did little against the Americans' superior Krags and the naval blockade further prevented arms shipments to the insurgents.

Strengths of the Counterinsurgents

1. *U.S. commanders' experience in irregular warfare in the Indian Wars.* Most of the U.S. military leaders in the Philippines, including Generals Otis, MacArthur, and Merritt,

[9] "[The U.S. Navy's] blockade of the archipelago effectively prevented Aguinaldo from receiving foreign arms shipments or moving supplies and reinforcements. Geography helped too: In the Philippines, there were no sanctuaries and no Ho Chi Minh trails to keep the guerrillas in business" (Boot, 2002).

[10] "Poorly trained and badly disciplined, composed of a mélange of volunteers, veterans of the Spanish colonial army, Katipuneros, and provincial forces, the Republican Army resembled a feudal levy more than a modern military organization" (Linn, 1989).

had served in the recently ended (1890) Indian Wars.[11] This rigorous experience with unconventional warfare gave commanders a number of skills and strategies with which to fight insurgents, which were a huge factor to winning in the Philippines. The most effective of these was the internment of civilians in concentration camps, originating from the successful tactic of restricting Indians to reservations, which was implemented toward the end of the conflict under Generals MacArthur and Chaffee.[12] The ostensible purpose of this was to separate civilians from insurgents and thus cut off their base of support within the population, particularly through the taxes insurgents levied as well as possible future recruitment and theft of property. Civilians were restricted from leaving the camps and a buffer zone between the camps and the outside was created in which everything was burned, so as to leave nothing for the insurgents. Another counterinsurgency tactic derived from the Indian Wars was the rationing of food among the local population, thereby severely restricting supply and removing any incentives for villagers to share their food with insurgents.

2. *Well-trained corps of soldiers.* Three separate types of the U.S. Army forces served in the Philippine War: the Regulars (about 65,000 soldiers at the start of the war), and the State Volunteers and U.S. Volunteers (about 35,000 soldiers combined). Many of the troops had previously served in Cuba and Puerto Rico, as well as in the Indian Wars, and had had some exposure to the types of climate and terrain they would encounter in the Philippines. They were also acquainted with the irregular warfare and contingency operations that would prove to be the bulk of operations in the Philippine campaign. Finally, U.S. troops were simply "better trained and disciplined"[13] than their Filipino counterparts who, for the most part, had no formal training in warfighting techniques.
3. *Development operations simultaneous to combat operations.* One of the most important steps taken by the Army to ensure popular support for its campaign was to pursue what became known as development or "contingency operations" at the same time as fighting the insurgents. From the outset, the Army was tasked with building (and sometimes staffing) schools, setting up governments within rural areas, and providing such amenities as modernized sanitation and communications systems, schools, hospitals, and roads. This helped sway a great number of civilians to the U.S. cause, especially after over three centuries under the Spanish, who had very little interest in providing for the populace, other than wealthy landowners. The original impetus for these contingency operations was President McKinley's order to defeat the insurgents in a humane manner, and many commanders took it upon themselves to give their soldiers

[11] "The Army's success may be ascribed in some degree to the invaluable experience its top commanders had gained in fighting Indians, the finest irregular warriors in the world. Out of 30 U.S. generals who served in the Philippines from 1898 to 1902, 26 had fought in the Indian Wars" (Boot, 2002).

[12] "Chaffee brought the Indian Wars with him to the Philippines and wanted to treat the recalcitrant Filipinos the way he had the Apaches in Arizona—by herding them onto reservations" (Carter, 1917, p. 5ff).

[13] Boot (2002).

a constant reminder of this.[14] The mission was formally described in General Orders (GOs) 43 and 40 of 1899 and 1900, respectively, which provided for municipal governments and police forces in localities where the Army was stationed. When given a choice between rebels who extorted money from villages or the U.S. Army, which made efforts to improve their quality of life, many Filipinos chose to side with the U.S. Army, thus providing the local support crucial to antiguerrilla operations.

4. *Adaptation of tactics.* As a result of the geographical and cultural diversity of the Philippines, U.S. Army commanders quickly learned to structure tactics in each theater according to individual conditions. In dealing with the Filipino population, commanders took note of the fact that some were inherently more receptive to the U.S. cause and had to be treated differently. They therefore created appropriate incentives to induce different populations to cooperate, such as imposing stricter punishments to make it unpopular to side with the rebels.
5. *General Order 100.* GO 100, established in 1863, was the code of warfare in the U.S. Civil War and, in the Philippines, it stipulated rewards for those who cooperated with U.S. forces and harsh punishments for guerrillas. After General MacArthur's appointment, the order was stressed to commanders to enforce a strong and very visible standard of justice in the Philippines. Besides informing the commanders, a larger effect of GO 100 was that this standard of justice became known to the Filipinos: Collaboration with the guerrillas, even indirect, would be punished severely.
6. *Development of Filipino auxiliary armed forces and police.* Along with the contingency operations the United States conducted, the Army also trained two distinct indigenous armed forces in the Philippines (the Constabulary and the Scouts) and a number of rural police forces. The Constabulary was semi-autonomous paramilitary forces set up to fight insurgents, a shining example of which was the contingent on the island of Negros led by Brigadier General James F. Smith.[15] The Scouts were a force originally organized by Lieutenant Matthew Batson in Pampanga Province which, because of its high rate of success at countering guerrilla activity, eventually grew into a military auxiliary corps of over 15,000.[16] In fact, the Scouts, under the direction of Major Frederick Funston, were responsible for the capture of Aguinaldo in North Luzon in February 1901, a major turning point in the war. In raising these forces, the Army handed over some of the responsibility for the security of the islands into the hands of the Filipinos. The effect of this was twofold: Filipinos were often more respected by the population and thus more effective at countering rebel influence, neutralizing insurgents and gathering intelligence. At the same time, the creation of a local Filipino force countered the

[14] "The Army was not only to suppress terrorism, guerrilla warfare, and brigandage but to prepare the Philippines for colonial government; moreover, this must be accomplished in such a way that the Filipinos would be docile, obedient, and grateful subjects" (Linn, 1989).

[15] "The battalion was to prove among the most successful scout-police forces the U.S. Army raised, without a single deserter or lost rifle in its entire existence" (Linn, 2000).

[16] Linn (2000).

image that the Americans were ruthless colonizers. Along with other steps toward self-government, the creation of indigenous forces allowed the Filipinos to largely police themselves.

Weaknesses of the Counterinsurgents

1. *Very narrow top-down view of insurgency in the early stages.* As commander of U.S. forces in the Philippines for the first stage of the conflict, General Otis assumed a wholly unrealistic view of rebel demographics and ideology. After talking to a number of wealthy Filipinos from his outpost in Manila, Otis decided that "Aguinaldo and the revolutionary forces represented either Tagalog despotism or lower-class anarchy and that a firm policy toward them would secure support from 'men of property.'"[17] This notion is quickly refuted by the simple facts that Aguinaldo was himself a *principale* and that a large segment of this group supported him throughout most of the conflict. Nonetheless, it persisted in some commanders' thinking even after the broader-minded General MacArthur succeeded Otis, by which time the conflict had progressed into all-out guerrilla warfare.
2. *Occasional brutality.* The flip side of soldiers' experience in asymmetric warfare during the Indian Wars is that they were also informed by the racist rhetoric behind the conflict.[18] Thus, when they landed in the Philippines, many brought these attitudes along with them, making the war unpopular both at home and among indigenous Filipinos. Previously mentioned tactics, such as internment of civilians in concentration camps and food deprivation (and destruction of property), while effective, reduced confidence in U.S. troops and took a high toll on the population in the form of malnourishment. Aguinaldo attempted to exploit this to his own ends in thinking that the unpopularity of the war would swing the vote in the United States toward the Democrats in 1900 and end the U.S. occupation. The final phase of the war in 1901–1902 saw a number of violent counterinsurgency operations unacceptable by modern standards; the retaliation on the island of Samar to the murder of 48 U.S. soldiers in Balangiga was particularly brutal. Summary executions were common, usually in retribution for guerrilla violence, and torture methods, such as the water cure, became commonplace as a method for obtaining information.
3. *Harsh climate and terrain and lack of supplies.* The Army found highly unwelcoming conditions once they landed on Luzon and these would remain a significant problem as fighting spread beyond the island. The whole force suffered from a lack of appro-

[17] Linn (1989).

[18] "Some [U.S. commanders] had taken part in the massacre at Wounded Knee. It was easy for such commanders to order similar tactics in the Philippines, particularly when faced with the frustrations of guerrilla warfare. And the men in their command, many of whom were themselves descendants of old Indian fighters, carried out these orders with amazing . . . alacrity" (Miller, 1982).

priate equipment, weapons, and food throughout the conflict. The geography of the country did not help either, as soldiers often had to march over mountains only to find themselves in the middle of a dense jungle once they had finished their descent. Consequently, hundreds became sick and died; many more suffered extreme exhaustion and depression, leaving a large part of the Army generally demoralized.[19] Another problem caused by difficult terrain was lack of communication among the various fronts during the war, often leading to decisions by commanders that contradicted those of others and were ultimately counterproductive.

Conclusions

As the U.S. sole unequivocal counterinsurgency success, the war in the Philippines holds a number of lessons for future operations. The most important factor in winning the war was likely the attention devoted to gaining the support of the populace. Through programs such as the creation of strong municipal governments, schools, and infrastructure within the country, the United States was able to ultimately convince the majority of the population, many of whom were potential enemy combatants, not to side with the insurgents. The commander most wholeheartedly committed to this was General MacArthur, who took the most significant steps to gaining the support of both the *ilustrado* educated class and the *principale* landowners. Under MacArthur, the auxiliary armed forces (Scouts and Constabulary) programs really took off, providing the United States with heretofore impossible-to-attain intelligence and a more effective way to control and protect the indigenous population. Such programs as concentration camps and food deprivation, while controversial, helped to cut off insurgents from bases of support in the civilian population and harsh punishments for collaboration were made visible to the Filipinos in GO 100. The last and most brutal phase of the war, under General Chaffee, contributes the most to the view of the conflict as an unjustifiably violent one from the side of the Americans, and rightly so. Exasperated with the insurgents' guerrilla tactics and attempting to extricate U.S. troops from a war that had dragged on too long, commanders too often overstepped the bounds of President McKinley's mandate to "win the confidence, respect and admiration"[20] of the Filipinos. Although not all U.S. tactics could be justified, it remains a fact that, in less than four years, the U.S. Army was able to permanently crush a large insurgency in a country where it had previously had no experience.

---

[19] "Cholera, dysentery, malaria, venereal diseases, and sheer heat exhaustion ravaged the ranks, depleting some units of 60 percent of their strength" by only the spring of 1899 (Boot, 2002).

[20] McKinley (1898a, p. 859).

CHAPTER THREE

Algeria (1954–1962)

Peter Chalk

Introduction

Known as the "First Algerian War," the anticolonial struggle that was conducted by the *Front de Libération Nationale* (FLN) against France is often portrayed as a model of classic insurgent warfare that effectively combined guerrilla strategies with a highly brutal campaign of urban terrorism to discredit and ultimately destroy Paris's stated conviction of the need to hold onto its North African outpost. Indeed, the lessons derived from the Algerian experience were to greatly influence many later ethnonationalist insurgent campaigns, including those conducted by the Palestine Liberation Organization (PLO) and the African National Congress (ANC).[1] (See Figure 3.1 for a map of Algeria.)

This chapter provides an overview of the Algerian Civil War between 1954 and 1962. It begins with a discussion of the conflict, then examines some of the key weaknesses and strengths exhibited by rebel and government forces on the ground. The case study highlights the crucial political dimension of counterinsurgency—namely, that although forceful military actions can work to seriously degrade the operational capabilities of an insurgent movement, this alone will be insufficient to secure a complete victory in the absence of a favorable political contex.

Origins and Characteristics of the Insurgency

The FLN launched its insurgency in the early hours of November 1, 1954, with a force of 300 *masquisards* (fighters). These rebels, who were armed with an assortment of antiquated weapons left over from World War II, conducted a series of attacks against security force installations, police posts, warehouses, communication facilities, and public utilities. Justifying the assaults in the name of a legitimate struggle of independence from the French and colonial settlers (known as *colons*), the FLN broadcast a proclamation from Cairo urging all Muslims to

[1] Hoffman (1998, pp. 60–61).

Figure 3.1
Algeria

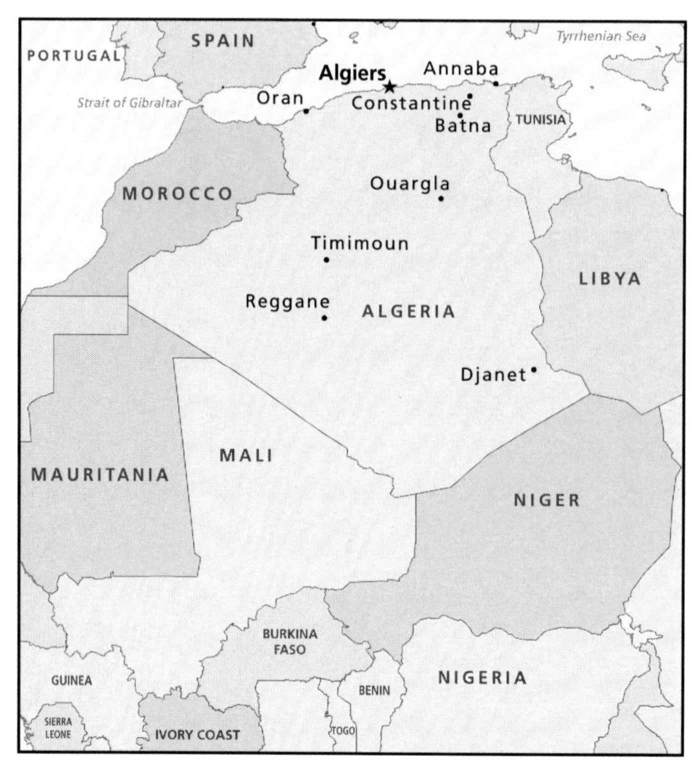

RAND *OP185-3.1*

join in a national struggle for the "restoration of the Algerian state, sovereign, democratic, and social within the framework of the principles of Islam."[2]

The basic thrust of the FLN's strategy during the initial stages of the war focused on creating resistance groups and cells whose main task was to recruit new members and impress a pro-independence mindset on the Algerian Muslim community. Subsequently, however, a more explicit tactical use of urban-based terrorism became evident, which was essentially aimed at two main objectives: first, to provoke an overreaction on the part of the security forces that could then be exploited to internationally highlight the Algerian struggle for independence; second, to drive a wedge between the local population and the colonial administration.

The FLN's shift to urban-based modalities became readily apparent between the end of 1957 and the beginning of 1958 when, during the course of four months, the organization conducted a range of operations in the heart of Algiers' European sector. Notable incidents included two waves of synchronized bombings that specifically targeted civilian-centric "soft" venues (collectively killing 18 and injuring over 140);[3] the assassination of the city's mayor,

[2] Wikipedia (2006).

[3] The bombings were conducted in two major waves—one in September 1956 and one in January 1957. Targets included the Milk Bar (a seaside venue especially favored by the Colons); the Cafeteria (one of the main haunts frequented by Euro-

Amédée Froger; an attempted massacre of the cortege accompanying the stricken leader's subsequent funeral procession; killings of several other high-ranking officials in the colonial administration; and the staging of a general strike that seriously disrupted the capital's postal, telegraph, and railway services for a number of days.[4]

In response to these actions, the French military—which had considerable autonomy on account of the highly fractured nature of the country's political establishment at this time[5]—called out the elite 10th Parachute Division, granting it full authority to do whatever was necessary to restore order in the capital. In what subsequently became known as the Battle of Algiers, the decision proved to be one of great import. Rationalizing that extreme circumstances warranted extreme countermeasures, the unit's commander-in-chief, General Jacques Massau, authorized wholesale roundups of entire neighborhoods (enacted under a system of quadrillage in which the city was divided into controlled "squares"—each one conforming to a regional command) in addition to extrajudicial preemptive detentions of FLN suspects.[6]

Although Massau's actions were instrumental in crushing the FLN's terror-based operations, they elicited widespread international debate and consternation. Just as significantly, they fostered increased popular sympathy for the insurgents—who, despite carrying out bloody atrocities of their own, were increasingly viewed as "defenders of the people's rights"—as well as polarized and undercut domestic public opinion and support in metropolitan France itself.[7] Growing domestic and overseas pressure for a definitive end to the conflict eventually forced French President Charles de Gaulle, who had come to power in 1958 on a distinct pro-French Algerian policy, to reverse his stance and concede to a negotiating stance that explicitly recognized the possibility of colonial self-determination and majority rule.[8]

The president's U-turn sparked a major insurrection among the colons in Algeria who, in conjunction with hardline elements in the army who were adamant that France's earlier ignominious retreat from Indochina should not be repeated in North Africa, established the *Organisation de l'Armée Secrète* (OAS). An overtly terrorist entity, this organization committed itself to an open-ended agenda of political violence that was intended to provoke a leadership crisis in Paris (through the assassination of de Gaulle); unleash a civil war against the metropolitan government, police, and army; and ignite an ethnic war against Muslims.[9] Despite carrying out a wave of highly bloody attacks that, at its height, averaged 120 bombings a day, the OAS rebellion was contained (largely because the bulk of the military remained loyal to de Gaulle), allowing negotiations to take place between the FLN and Paris that culminated with the

---

pean students), the Coq Hardi (a popular brasserie) and the downtown Air France passenger terminus (Hoffman, 1998, p. 62; Horne (1977, pp. 185–186, 192).

[4] Horne (1977, pp. 183–194).

[5] The French political system of the 1950s accorded considerable authority to a parliament that was so fractured it could not produce a national leader sufficiently empowered to make decisive strategic decisions on Algeria policy.

[6] Horne (1977, pp. 198–199); Hoffman (1998, pp. 62–63); Martin (2005, p. 53).

[7] Hoffman (1998, pp. 63–64); Martin (2005, p. 53); Horne (1977, Chapter Eleven).

[8] Wikipedia (2006); Martin (2005, pp. 53–54); Hoffman (1998, pp. 63–64).

[9] Martin (2005, p. 54); Thackrah (1987, p. 174).

signing of the Evian Agreements on March 19, 1962. These provided for a cease-fire; the granting of a full range of civil, political, economic, and cultural rights for all Algerians; and the holding of a popular referendum to decide whether the territory should remain a constitutional component of France or become a sovereign state in its own right.[10] Although the OAS attempted to destroy the accord through a last-ditch campaign of urban terrorism, a vote on the future status of the colony was taken on July 1, 1962.[11] This subsequently returned a nearly unanimous[12] result in favor of independence, which was duly conferred to Algeria two days later.[13]

The war's overall toll was immense. During the course of eight years, an estimated 1.5 million Algerians either disappeared or were killed. In addition, over two million residents were dislocated from their homes and a third of the country's economic infrastructure destroyed. French military losses amounted to 18,000 dead (6,000 from noncombat-related causes) and 65,000 wounded, with European-descended civilian casualties exceeding 10,000 (including 3,000 fatalities) in 42,000 recorded terrorist incidents.[14]

Strengths of the Insurgents

The FLN's main strength revolved around its expeditious use of terror to alter the political context of the Algerian conflict. As noted in the preceding section, the adoption of this operational tactic was primarily designed to provoke overreaction on the part of the French security forces as well as to garner national and international support for the group's independence agenda.[15] In both endeavors, the FLN was remarkably successful and this remains perhaps the most enduring legacy of its insurgency.

The chief architect of the insurgent's terror strategy was Ramdane Abane, the FLN's top theoretician. Entirely self-taught, he was as wholly unsentimental as he was ruthless, retaining a complete and unalterable faith in the utility of civilian-centric violence. Many of Abane's views reflected the thinking of Carlos Marighela, whose "Minimanual of the Urban Guerrilla" is perhaps the best-known revolutionary treatise laying out the reasoning behind the use of terrorist tactics:

---

10 For a detailed account of the Evian Agreements, see Horne (1977, Chapter Eleven).

11 The OAS eventually signed a peace treaty with the FLN on June 17, 1962.

12 Some six million out of a total population of 6.5 million voted in the referendum.

13 Wikipedia (2006).

14 Wikipedia (2006).

15 The FLN's adoption of civilian-centric, urban-based modus operandi was also instrumental in conceptualizing the first philosophical efforts to justify the tactical resort to terrorism. The ideas of Frantz Fanon and Jean-Paul Sartre, both of whom advocated the systematic use of violence as a "cleansing" force that could free an individual from feelings of inferiority, despair, and inaction, were directly influenced by the Algerian War and represent, perhaps, the clearest attempts to exonerate terror as an all-embracing end in itself rather than merely a functional means to a particular objective. See Wardlaw (1982, p. 41).

> From the moment a large proportion of the population begin to take his activities seriously, his success is assured. The Government can only intensify its repression, thus making the life of its citizens harder than ever: homes will be broken into, police searches organized, innocent people arrested, and communications; police terror will become the order of the day, and there will be more and more political murders—in short a massive political persecution . . . the political situation of the country will become a military situation.[16]

Abane applied this logic to the Algerian context with dispassionate efficiency between 1957 and 1958, pointedly declaring that for every FLN fighter executed, a hundred Frenchmen would meet a similar fate. The taunt, which was subsequently mirrored in the unleashing of an intensive campaign of urban bombings in the heart of Algiers, was unambiguous and explicitly aimed at goading the authorities into the type of overreaction that could be readily leveraged to alienate the local population from the colonial administration. Just as important, Abane appreciated the role of terror in bringing an otherwise "silent" cause to the attention of the world community. This became clearly evident in the now infamous Directive Nine, in which he rhetorically asks: "Is it preferable for our cause to kill ten enemies in an *oued* [dry river bed] of Telergma when no one will talk of it or a single man in Algiers which will be noted the next day by the American press?"[17]

The FLN's adoption of terror proved decisive. Paris was, indeed, provoked into initiating a draconian, highly brutal response (discussed below) that completely alienated the native population, destroyed the middle ground of Muslim political compromise, and drove formerly passive Algerians, if not directly into the ranks of the FLN, at least away from the colonial administration. The security force's actions also served to disgust domestic public opinion in France, which, in turn, significantly undercut popular support for continuing with the war.[18]

Just as important, from the end of 1957, the conflict in Algeria became a notable feature on the international diplomatic and political landscape. Not only was the conflict hotly debated in the United Nations (UN) Security Council and General Assembly, but several leading Arab states also declared unambiguous support for the FLN cause in what was effectively the first concerted show of Middle Eastern unity on a global issue. In addition, there was growing awareness of the war in both Britain and the United States, with Washington, in particular, expressing concern that military equipment supplied to France for North Atlantic Treaty Organizations (NATO) purposes (notably helicopters) was being misdirected and used to quell the Algerian insurgency.[19] As Horne sums up:

> Disheartening to France as it was encouraging to the FLN in their campaign for "internationalization" was the growing awareness of the war in both the United States and Britain. Inevitably this brought hostility to France's role in it. As in France, the Battle of Algiers

[16] Marighela (1982, p. 37).

[17] Abane, quoted in Gaucher (1965, p. 262).

[18] Hoffman (1998, pp. 63–64); Martin (2005, p. 53); Horne (1977, Chapter Eleven).

[19] Wikipedia (2006).

> had done much to publicize the war, [which was now being echoed] in Anglo-Saxon . . . opinion.[20]

Weaknesses of the Insurgents

One of the main weaknesses of the insurgents, particularly in the early stages of the conflict, was a miscalculation that their rebellion would be actively supported by the civilian population. There were at least four reasons why the FLN believed that local Muslims would readily endorse its call to arms: First, the indigenous population was deprived of all political rights; second, virtually the entire Algerian economy was in the hands of the French or metropolitan settlers; third, many veterans who had fought with great bravery against Nazi Germany during World War II and subsequently alongside France in Indochina were dismayed at the lack of recognition they garnered from Paris, particularly the colonial government's refusal to grant them equal status, much less citizenship.

This assessment, however, quickly proved erroneous. Although national sentiment was not opposed to the attainment of sovereign independence per se, it was not especially radical and certainly far less so than in Morocco and Tunisia, both of which gained their independence at roughly the same time. Part of the reason for this derived from the support Paris garnered from notorious Muslim tribal chiefs and elitist judges, clerics, and bureaucrats whose interests were served by the French presence. In addition, although colonial rule did not fundamentally improve the political rights of the indigenous population, it did help to improve their socioeconomic condition through the provision of schooling and housing and the eradication of disease. Just as important, there was no concerted attempt by the colonial power to eradicate local culture and traditions, which allowed Algerians to assimilate French norms and customs without forsaking their own. Finally, there existed a strong middle ground of political compromise that was confident that a degree of genuine autonomy could be achieved through a peaceful, democratic devolution of power.[21]

The FLN's misreading of grassroots support for independence was particularly significant as it initially caused the group to substantially overestimate its chances for success on the battlefield. Indeed, the movement's leaders had concluded early on that France was a "military paper tiger"—a perception that was largely derived from a series of defeats inflicted on the colonial power in Indochina, Tunisia, and Morocco (all of which preceded the Algerian insurgency)—and would certainly be no match for an insurgency with a strong popular base. As Hoffman notes, however, two years after the commencement of hostilities, the FLN had precious few tangible efforts to show for its efforts and there was certainly no sign of an imminent French withdrawal from the territory.[22]

---

[20] Horne (1977, p. 243).

[21] Hoffman (1998, p. 61); Martin (2005, pp. 52–53).

[22] Hoffman (1998, p. 61).

Strengths of the Counterinsurgents

Although the French were ultimately forced to withdraw from Algeria, some aspects of its campaign against the FLN are worthy of consideration in terms of effective COIN. Noteworthy was the institution of a two-pronged doctrine of pacification that was pursued following the (draconian) restoration of order in Algiers and that was aimed at (1) obtaining the support of the population and (2) "starving" the FLN of vital external support and internal territorial control. To accomplish the first task, Paris moved to provide humanitarian assistance to local Muslim communities, pledged full protection for those who sided with the army, and, through the notion of "association," guaranteed that the colony would be recognized as a constituent part (though not integral component) of the French Republic[23] with concomitant rights guaranteed for all its citizens.[24]

Simultaneously, the French army moved to stop the external flow of materiel and personnel support to the FLN by reestablishing control over Algeria's land borders. To this end, an electrified barbed-wire fence complete with minefields, radars, and patrol zones for armed elements was constructed along the colony's borders with Morocco and Tunisia. The barrier was intended to act as a fishnet for the frontier regions that security force interdiction units could then use to both round up rebels who had covertly crossed into Algeria as well as identify the routes they had taken.[25]

The ultimate goal of pacification was to politically and operationally asphyxiate the FLN in order to reestablish a secure environment for reunification without exposing the people to excessive risk. French Army Colonel Gilles Martin explains how the concept was put into practice:

> To achieve these goals, the Army replaced a civilian administration unable to act in secure areas. It took over the management of schools, clinics, road maintenance, the water supply and so on. To help administer these functions, the Army divided Algeria into a "grid" of regions, sectors and sub-sectors. At the lowest level, an infantry company controlled a few villages and a couple thousand inhabitants. The same soldiers who used shovels, first-aid kits, and schoolbooks reinforced security, administered the population and fought [the rebels].[26]

Pacification worked well in several respects. The strategy, which necessitated extensive and sustained contact between army units and their assigned villages, helped to build a measure of trust within elements of the local population. Although this by no means neutralized the lingering effects and trauma caused by Massau's brutal campaign in Algiers, in a number of instances, communitarian bonds were established and leveraged to create self-defense units. Known as *harkas*, these forces operated in tandem with the military and were effectively used

---

[23] It should be noted that this provision arguably undercut Paris's hearts and minds strategy, as it effectively forced Algerians to accept permanent inferiority as a department of France.

[24] Martin (2005, pp. 55–56). See also Galula (2006).

[25] Martin (2005, p. 56).

[26] Martin (2005, pp. 55–56).

to seek out and destroy rebel strongholds. Equally, the use of border barriers was so efficient that it made external infiltration more or less impossible, prompting FLN cadres across the Maghreb to deliberately abandon their home-based comrades. Reflecting these successes, by 1960, the insurgents had no more than 5,000 members, no firm area from which to conduct and plan offensive attacks, and no real objective beyond survival.[27]

Weaknesses of the Counterinsurgents

By far the greatest weakness (and mistake) exhibited by the counterinsurgents was how they reacted to the FLN's adoption of an urban-based campaign of terrorism. As suggested above, the response by the security forces was as absolute as it was brutal. Indeed, the various methods employed by General Massau to restore order and control in Algiers effectively amounted to an institutionalized regime of torture. Interrogation techniques were extremely harsh, embracing electrocution, simulated drowning, and directed abuse specifically aimed at degrading human dignity. Detainees who refused to talk were eliminated and, together with those who simply died during questioning, disposed of in "cleanup" operations that became so prevalent they earned the slang expression "work in the woods."[28] Overall, it is thought that almost 40 percent of the *Casbah* (Arab quarter) were arrested or detained during the course the four months, some 3,000 of whom were never heard from again.[29]

That these tactics were effective in crushing the FLN's operational capacity there is no doubt. However, France was unable to capitalize on the group's battlefield weakness, largely because the government's use of ruthless and repressive tactics during the Battle of Algiers had fatally exposed the bankruptcy of the central government's rule. Not only had indigenous Muslims been irreconcilably estranged, the metropolitan population had become totally disillusioned with the idea of an Algerie Française and international opinion was squarely behind decolonization. Thus, although Massau claimed that the Parachute Division's actions were justified in smashing the FLN's urban campaign of terror, it is evident that they came at enormous political cost.[30] As Paul Teitgen, the Secretary-General of the Algerian Governor-General at the Prefecture remarked, "All right, Massau won the Battle for Algiers; but that meant losing the war."[31]

The reality of the situation was quickly recognized by de Gaulle, who within two years of assuming the presidency, had come to accept that retaining control over the North African territory was no longer tenable—even in the face of vigorous opposition to a French withdrawal by both colonial settlers as well as hard-line elements within the French army.[32] Accepting the

[27] Martin (2005, pp. 55–56); Wikipedia (2006).

[28] Horne (1977, pp. 198–206); Wikipedia (2006); Hoffman (1998, pp. 62–63); Martin (2005, p. 53).

[29] Wikipedia (2006); Rosie (1986, p. 48); Horne (1977, pp. 201–202).

[30] Hoffman (1998, p. 64).

[31] Teitgen, in Horne (1977, p. 207).

[32] Wikipedia (2006); Horne (1977, pp. 234–250); Martin (2005, pp. 53–54).

1962 Evian Agreements with characteristic cynical realism, he addressed his cabinet the day after their signing, declaring, "It's an honorable exit. It's not necessary to write an epilogue on what has just been done, or not done. . . . That the application of the Agreements will be capricious is certain. . . . As for France, it will now be necessary for her to interest herself in something else."[33]

Conclusions

It is a widely accepted truism that France won the military battle in Algeria but lost the political war. Goaded into a "no holds barred" policy of eradication by the FLN's adoption of terrorism, the case study is a sad but necessary illustration of what can occur when a Western democracy relinquishes its authority and responsibility and allows the military to conduct a COIN campaign unfettered by the rule of law. As Hoffman observes, the Battle of Algiers remains perhaps the most significant episode in the FLN's insurgency in that it succeeded in removing any semblance of moral legitimacy from Paris, while simultaneously exposing the rebels to favorable world attention—just as Abane had calculated.[34]

A full and complete victory for the French was essentially impossible under such circumstances. Irrespective of its military standing, by the end of the 1950s, the political initiative lay firmly with the FLN. The FLN's government in exile—the *Gouvernement Provisoire de la République Algérienne* (GPRA, which was established in Tunis in late 1958)—had been recognized by Morocco, Tunisia, and the Soviet Union as well as several other Arab, Asian, African, and East European states; labor unions in Britain and the United States were openly calling for Algerian Muslim self-determination; and in France a growing antiwar movement had begun to take root.

The political dimension of the Algerian conflict was a lesson that was to inform the thinking and tactics of numerous national liberation movements across the world. Yassir Arafat, for example, explicitly singled out the seminal influence that the FLN experience had on the PLO's own struggle against Israel—one that was to be reflected in the intensive diplomatic agenda pursued by the late Palestinian leader throughout the 1990s and 2000s.[35] Nelson Mandela similarly identified the civil war as one that was to have a major bearing on the ANC's decades-long effort to end white minority rule in South Africa. Observing that the North African colonial conflict was the closest model to the situation confronting his own movement, he readily appreciated the crucial significance of the nonmilitary sphere on successful pro-independence struggles, taking away the chief lesson that "international [and domestic] opinion . . . is sometimes worth more than a fleet of jet fighters."[36]

---

[33] de Gaulle, in Horne (1977, p. 523).

[34] Hoffman (1998, p. 64).

[35] Hoffman (1998, p. 60). See also Hart (1994, pp. 112–113); Cooley and March (1973, p. 91); Hirst (1977, pp. 273, 276, 306–307); O'Balance (1974, pp. 23, 26); Schiff and Rothstein (1972, pp. 8, 60).

[36] Mandela (1994, p. 355); Hoffman (1998, p. 61).

CHAPTER FOUR

Vietnam (1959–1972)

Lesley Anne Warner

Introduction

This chapter will discuss the insurgency in Vietnam, starting with a brief background on the origins of the conflict. Special attention will be paid to the years 1959–1965 and 1968–1972, as these were the years in which the United States made several attempts at pacification in South Vietnam, which met with mixed success. For the purpose of this paper, North Vietnam is discussed in conjunction with the insurgent forces because that country supported and directed the insurgency in the south and recognized the inseparability of the political and military aspects of the war. The part of the war discussed in this paper had three phases: the beginning of the insurgency (1959–1963), post-Diem instability and the arrival of U.S. combat troops (1963–1968), and the pacification era (1968–1972). (See Figure 4.1 for a map of Vietnam.)

Origins and Characteristics of the Insurgency

U.S. involvement in Indochina began during World War II, when the Office of Strategic Services (OSS) aided the Vietnamese resistance to Japanese occupation. President Harry S. Truman subsequently authorized $3 billion in aid to the French to regain their control over the country after the war.[1] It was a question of leverage, not interest in the region, that spurred an increased U.S. commitment: U.S. aid for the war in Indochina in exchange for French involvement in NATO.

Following the French surrender at the battle of Dien Bien Phu in 1954, Vietnam was divided at the 17th Parallel into North and South Vietnam and was to be reunified with an election in 1956, as prescribed by the Geneva Accords. By this time, the perception had changed in U.S. policymaking circles regarding the importance of Southeast Asia; North Vietnam was perceived as a fallen domino in the Cold War, and South Vietnam was seen as a domino at risk.

---

1 Karnow (1997, p. 192).

Figure 4.1
Vietnam

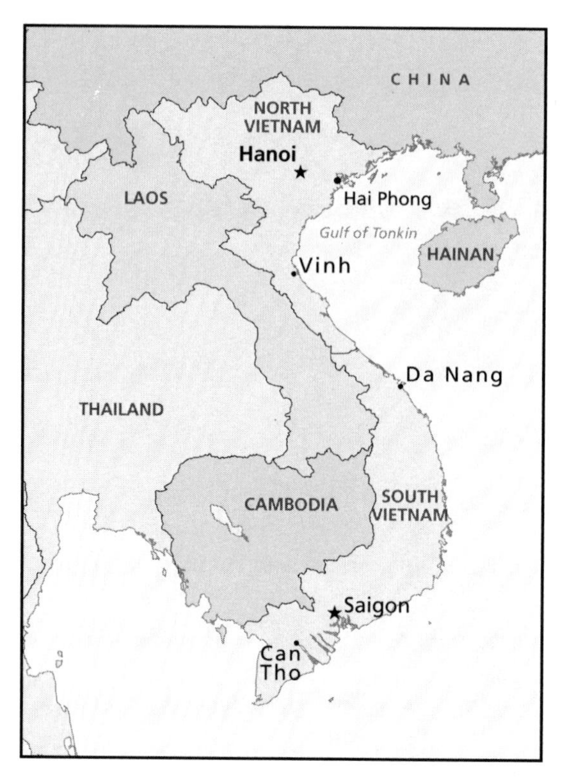

RAND *OP185-4.1*

Phase I: The Beginning of the Insurgency (1959–1963)

The insurgency in South Vietnam evolved in 1959 from social cleavages as well as from North Vietnam's desire for reunification. At the beginning of the insurgency, South Vietnam was a country whose future had been placed in the hands of Ngo Dinh Diem, a U.S.-educated Catholic who was out of touch with the rural, predominantly Buddhist population. In his seven years in power, Diem eliminated village autonomy, instituted land reforms that increased the power of the landowning elite, and exacerbated social tensions by promoting to government posts family members and Catholics who were often tainted by association with French colonialism. When Diem refused to hold elections for reunification and began to have moderate success arresting his political opponents, many of them communists, North Vietnam decided to begin its reunification campaign. To avoid violating the Geneva Accords, it founded the National Liberation Front (NLF) in 1960 as a façade for its covert activities in South Vietnam.[2] In 1961, U.S. advisers were sent to South Vietnam to help combat the developing insurgency.

---

[2] In addition to members of the NLF, thousands of Viet Minh from the war against the French had remained in the south to aid the renewed war effort.

Phase II: Post-Diem Instability and Arrival of U.S. Combat Troops (1963–1968)

Nepotism, corruption, and Diem's further alienation of much of the population following the 1963 summer protests and monk self-immolations hastened the demise of his regime and he was overthrown and assassinated in November 1963. For the better part of the remainder of U.S. involvement in South Vietnam, the leadership of the country became a revolving door, further obstructing the goal of defeating the insurgency. North Vietnam, seeing the resulting chaos following the coup, increased its infiltration of the south.

In August 1964, on the assumption that two U.S. destroyers had been attacked in international waters in the Gulf of Tonkin, President Lyndon Johnson requested that Congress pass the Gulf of Tonkin Resolution—in effect, a blank check that facilitated U.S. escalation of the conflict. In 1965, the airfield at Pleiku was attacked, giving Johnson a pretext to send combat troops to Vietnam. When the troops arrived, the situation was such that the pacification that had taken place under Diem was put on the back burner, as the Americans and South Vietnamese struggled to stave off military defeat by main force Viet Cong and North Vietnamese military units.

Phase III: The Pacification Era (1968–1972)

The North Vietnamese and the Viet Cong launched the Tet Offensive in January 1968, the same year in which the U.S. troop level in South Vietnam peaked to just over 500,000. Tet was a watershed for both sides as, in the aftermath of Tet, the bulk of fighting shifted from the severely weakened Viet Cong to the North Vietnamese Army. On the U.S. side, public opinion in opposition to the war forced the government to seek disengagement from the conflict by increasing South Vietnam's ability to fight the war and thereby reducing the need for a U.S. combat presence, a process that was commonly referred to as "Vietnamization."

Because of the insurgents' weakness following Tet, the United States and South Vietnam were able to uncover much of the insurgent infrastructure and intensify pacification efforts with moderate success. Despite their fear of alienating powerful landowners, the South Vietnamese government enacted a land reform law designed to eliminate some of the population's grievances and co-opt one of the insurgent causes. Unfortunately, many of the reforms undertaken during this phase were enacted too late to be effective.

Strengths of the Insurgents

Throughout the conflict, the insurgents learned from other insurgencies and were adaptable in their techniques.[3] Because of their experience resisting the Chinese, Japanese, and French occupations, they had already learned which methods worked against a more powerful enemy. Therefore, when U.S. combat troops entered the conflict in the early 1960s, the insurgents were able to start from a tested base of knowledge built over the past 15 years. Many South Vietnamese cadres were brought north for training and education in 1954 and subsequently

[3] Ho Chi Minh lived in Moscow for two years studying how to organize a revolution and the North Vietnamese and Viet Cong learned from other insurgencies and followed Mao's example.

reinfiltrated, which added to that base of knowledge. The insurgents were also sensitive to the facts that the war and its tactics were constantly changing and that, to avoid defeat, they needed to change at an accelerated pace.[4]

Between 1954 and 1963, North Vietnam addressed the weaknesses in its own regime and consolidated power while carefully monitoring events in the south, fearing that if it were to act prematurely, it would destroy all it had gained during the period of internal stabilization following the French withdrawal.[5] The North Vietnamese respected Diem's nationalist credentials to the extent that he would prevent a large U.S. military presence, which is why they did not use force when Diem refused to hold elections in 1956.[6] They also maintained hope that his regime would collapse on its own as a result of its unpopularity and that they could quickly overrun the south before the Americans could react.[7] Despite these measures, North Vietnamese sources say that the insurgents took too long to transition from political struggle to armed struggle.[8]

To build a political and military base for guerrilla war and exploit the unrest resulting from Diem's inept governance, the North Vietnamese created the NLF in 1960, which was an umbrella organization consisting of the Viet Cong, the Cao Dai and Hoa Hao religious sects, and minority groups who felt alienated from the South Vietnamese government, such as the Montagnards and Khmers.[9] The NLF was directed from North Vietnam and led by Dang Lao Dong (Vietnam Workers' Party) members as well as South Vietnamese who had been trained in the north and reinfiltrated back to the south. The People's Revolutionary Party (PRP), created in 1962, served as the southern counterpart to the North Vietnamese Dang Lao Dong and its infrastructure operated at six levels: Central Office for South Vietnam (COSVN), region, province, district, village, and hamlet.[10]

The NLF structure was extremely decentralized, which contributed to the movement's success, as the village party committee had a great deal of autonomy in determining local policy.[11] Politically, the insurgents couched their appeals in terms of liberation from foreign oppression and rule by corrupt puppet leaders and made sure that their leadership included noncommunist figures to broaden the appeal of their cause. Overall, their appeal was one of

---

[4] "We must pay attention to the development of forms of war so that we can respond to the requirements of each period. When it is necessary, we must change in time outdated forms of warfare, taking up new ones which are more appropriate. . . . We should not apply old experiences mechanically, or reapply outmoded forms of warfare" (Vô, 1976, p. 8).

[5] McNamara, Blight, and Brigham (1999, p. 32).

[6] McNamara, Blight, and Brigham (1999, p. 329).

[7] As it turned out, the North Vietnamese had not prepared enough cadres to infiltrate the south because they had failed to anticipate the development of revolution (Pribbenow, 2002, p. 69).

[8] Pribbenow (2002, p. 117).

[9] Hunt (1995, p. 6).

[10] Hunt (1995, p. 110).

[11] In contrast, areas controlled by the centralized Saigon government were staffed by politically appointed outsiders, which made villagers feel that they had little influence in matters that concerned them and were not represented at even the lowest levels of government.

an alternative to the South Vietnamese government that would give them the opportunity to have a better life.

In terms of cadres assigned to each task, political activity exceeded military activity.[12] Three-man cells kept fighters disciplined, maintained the flow of information, and ensured that fighters were indoctrinated with the purpose of the struggle. Cadres were left in their home areas to work among their own people, which helped draw recruits from established social networks, exploited their familiarity with the villagers and the target area, and made use of preexisting trust.

Acting as a shadow government, the NLF sought to alleviate hardships by providing food, implementing land reform, lowering taxes, and tightening security in the countryside. It created mass associations at the village level, which were integral to the political struggle and to the involvement of the population in the movement and helped move their primary association away from their traditional family structures and closer to a communal organization. As in North Vietnam, the insurgents built free schools and health clinics and used face-to-face encounters with citizens and psychological operations (PSYOPs) through the press and radio to counteract U.S. leaflet campaigns. The insurgents also used discriminate terrorism to keep both committed and uncommitted villagers and cadres in line, knowing that the South Vietnamese government was too weak to protect them.

The insurgents attacked symbols of U.S. and South Vietnamese authority to highlight counterinsurgent inability to provide security for the population. Using propaganda, assassination, ambush, and other forms of terrorism and coercion, they intimidated enemies and encouraged fear and respect among those already committed to the cause. They also collected taxes at will and destroyed many main roads, making it difficult for the South Vietnamese government to control rural areas.

The insurgents sought sanctuary in foreign territories and used it for the transportation of weapons because the U.S. fear of escalation prevented the bombing of Laos and Cambodia until the final years of the war. This strategy allowed the insurgents to control their losses, dictate the pace and intensity of the war, and hold the strategic initiative. They knew that U.S. public opinion would not support a high-cost, protracted conflict, as was the case in France in the 1950s. In North Vietnam, Chinese manpower was used so that their own men could be available to fight in the south, which deterred the United States from bombarding close to the border with China.[13]

Weaknesses of the Insurgents

Although the insurgents were successful in surprising the counterinsurgents and in securing a psychological and political victory during Tet, the military operation was a failure because it devastated the Viet Cong manpower supply in such a way that, for the remainder of the war, the North Vietnamese troops bore the brunt of the fighting. Although Tet was considered a

---

[12] Pike (1986, p. 234).

[13] Schulzinger (1997, p. 210).

political victory for the insurgents, they lost many supporters as a result of their brutality in the villages they conquered.

After Tet, the insurgents continued to suffer from public relations errors because they had to resort to forced enlistment to replenish their supply of fighters. Because new recruits were not volunteers, they were not as committed to the cause as fully indoctrinated voluntary recruits had been. Additionally, because of population displacement as a result of Tet, the insurgents found it harder to depend on the population for supplies and taxes; thus, some areas were taxed harder than others, leading to additional alienation from the insurgent cause. Overall, the upheaval caused by Tet enabled the counterinsurgents to intensify pacification and made the contested population more inclined to support these efforts.[14]

In addition to these post-Tet difficulties, the insurgents had several ongoing weaknesses. The indoctrination and organizational style of the insurgents was such that they used a lot of paper in analyzing their progress and planning for the future, which created a lot of intelligence for the counterinsurgents once cadres and insurgent bases were captured.[15] Additionally, insurgent military and organizational techniques were slower to adopt a more ideal pace of progress, making the insurgents feel that they were not keeping up with the progress and adaptation of the counterinsurgents.[16] North Vietnamese sources acknowledge that the insurgents underestimated the effects of pacification and counterinsurgent capabilities and were thus slow in executing countermeasures.[17] Finally, the insurgent bases in liberated zones were small and isolated, which exacerbated their efforts to replenish manpower and equipment.

Weaknesses of the Counterinsurgents

As of 1962, the U.S. Army was doctrinally unprepared to act in its capacity as an adviser in counterinsurgency, which partially explains the difficulties the Americans and South Vietnamese had with pacification in the early 1960s. Although the Army Field Manual 100-5 discussed how to wage war against an enemy that used irregular tactics, it neglected to discuss good practices for training the forces of a host nation in this capacity.[18] That said, even if the United States had been equipped to advise Army of the Republic of Vietnam (ARVN), there is a good chance they still would have encountered difficulties, as ARVN was generally incompetent at both conventional and unconventional operations.

As a result of the U.S. predilection for fighting conventional wars, the Americans trained the South Vietnamese to counter a conventional invasion from the north while shortchanging the struggle against the insurgents in the south.[19] Additionally, the Army Chief of Staff, when

---

[14] Cassidy (2004).

[15] Blaufarb (1977, p. 214).

[16] Pribbenow (2002, p. 92).

[17] Pribbenow (2002, p. 237).

[18] Hunt (1995, p. 19).

[19] Hunt (1995, p. 13).

asked to train soldiers for counterinsurgency operations, allegedly insisted that "any good soldier could handle guerrillas."[20] Thus, instead of soldiers being trained thoroughly in the art of counterinsurgency, they were minimally trained for COIN as an added duty. Consequently, there was a significant time lapse of several years before the problem was correctly identified *and* the solutions correctly applied.[21]

The insurgents did not have an end date on their tour of duty and were committed to the war for the duration of the conflict, unlike U.S. soldiers, who served 12-month tours of duty. The short duration of their tours of duty and the short-term mindset meant that by the time they understood what techniques were required to combat the insurgency, they were no longer on duty.[22] With the mindset that Vietnam was just another small stepping stone in the Cold War, officers were rotated every six months to broaden their combat experience, which prevented them from gaining an intimate knowledge of their areas of operation.

Technology dictated the type of war that was fought by the counterinsurgents and had the unintended effect of working against the goals of pacification. The United States relied heavily on airpower and firepower because it was averse to heavy casualties and possessed an overwhelming technological advantage. This strategy often harmed civilians, especially "harassment and interdiction" fire, where the army fired at an unseen enemy with no particular target in mind. Large-scale operations were preceded by heavy bombing, alerting insurgents that counterinsurgent forces would soon be entering an area.

In addition to radicalizing the population, the counterinsurgent predisposition to use technology foiled one of the key elements of pacification—the provision of security for the population in exchange for intelligence. Sir Robert Thompson, a British counterinsurgency expert renowned for his involvement in pacification in Malaya, believed that the United States relied too heavily on helicopters and were never mobile on their feet, which hindered their ability to gather intelligence from the population. Conversely, the Viet Cong were able to control contested populations because they lived among the population and discredited South Vietnamese government officials in Saigon who were rarely physically or psychologically proximate to the population. Viet Cong proximity to the population also made it more complicated for the counterinsurgents to discern who the enemy was, coupled with the fact that in many cases, the Viet Cong *were* the population.

As of March 1964, it was estimated that the Viet Cong controlled 15 percent of the people and 30 percent of the territory in South Vietnam.[23] Because of the rapidly deteriorating political and military situation in South Vietnam, the counterinsurgents were forced to focus on the military dimension of the insurgency with little regard for the inseparability of the political and military in insurgent strategy.[24] In effect, the counterinsurgents were fighting two utterly

[20] Hunt (1995, p. 19).

[21] The John F. Kennedy Special Warfare Center and School at Fort Bragg, North Carolina, was established and developed as a lesson learned from the Vietnam War to address the need for specialized training for small wars.

[22] Schulzinger (1997, p. 196).

[23] Hunt (1995, p. 25).

[24] Throughout the conflict, the United States feared a Korea-type invasion from North Vietnam and there is much controversy over whether Military Assistance Command Vietnam (MACV) Commander, General William C. Westmoreland,

disconnected wars—a problem derivative of the lack of unity of command. The breakdown of communication between diplomats and military advisers contributed to the persistent lack of coordination. Once the United States became aware that the war could not be won unless it addressed both military and political issues, a realization embodied by the CORDS program, the situation had deteriorated to the point that the reforms could not have a sufficient impact.

The police, often the root of intelligence collection, provision of local security, and unearthing insurgent infrastructure, were understaffed, underfunded, and poorly trained. The National Police Force had significant trouble recruiting suitable candidates because preference went to ARVN for males between the ages of 18 and 28 and the recruits they did manage to attract were not paid very well.[25] Consequently, the police were often involved in corrupt activities that compromised the success of pacification.

Although the Phoenix Program, which will later be described in detail, was successful at gathering intelligence from prisoner interviews in an efficient manner, it was plagued by several weaknesses, as well as accusations of being an assassination program. Quotas encouraged the South Vietnamese to make arrests in retribution or simply to achieve the suggested number of insurgents captured and neutralized, and quotas were later eliminated because of this problem. Additionally, the legal system was not advanced enough to process the detainees, making the prisons breeding grounds and indoctrination centers for villagers who had previously been indifferent.[26] The counterinsurgents were eventually able to establish a system for processing prisoners to mitigate the program's weaknesses.

Much of the problem with pacification was the unreliability of the Saigon government. Although there was significant debate among U.S. officials as early as 1955 as to whether Diem was the right man for the job, the United States sought no alternative to his leadership until 1963, even though his government lacked authority and popularity with a majority of the population. Like the leaders who followed him, Diem's greatest concern was staying in power and he was consequently unable to deal with the insurgency. Despite the leader's inability and sometimes outright refusal to implement political reforms that would have abetted pacification, the United States did not have the option of completely cutting aid to South Vietnam for fear of further weakening the country's government.[27]

Since foreign embassy personnel were discouraged from traveling outside Saigon, the Americans developed an unhealthy reliance on governmental sources, which were not always trustworthy but reduced the perceived need to gather intelligence from the population in the countryside, further allowing counterinsurgents to maintain their distance.

---

made the correct decision to pursue a primarily conventional war of attrition when U.S. combat troops arrived in 1965. The author believes that it might have been the wrong decision to shortchange the political and unconventional aspects of the war at this time, as the methods of fighting the war should not have been seen as zero-sum and should have been addressed in terms of long-term effectiveness.

[25] Hunt (1995, p. 104).

[26] Tovo (2005, p. 12).

[27] Hunt (1995, p. 14).

Strengths of the Counterinsurgents

Throughout the course of the war, several specific programs were designed and implemented with the goal of improving pacification efforts. In 1966, a Revolutionary Development cadre program was established to mimic the organization of the NLF. In this program, highly trained and motivated men were sent to work in villages to ameliorate conditions for villagers and establish rapport between the governors and the governed. Some units even used the same hit-and-run tactics of the insurgents against conventional enemy forces. U.S. presence was often crucial to the success of these programs because the populations in question, especially ethnic minorities, tended to be suspicious of the intentions of the South Vietnamese government.[28] These particular programs designed to mobilize the population had mixed outcomes. On one hand, it showed that the United States knew enough about the cohesiveness of certain populations to build civil defense units around them. On the other, the ARVN units that were supposed to come to the aid of village defense units were often slow in responding, if they responded at all, which diminished the villagers' view that the Government of South Vietnam (GVN) could provide for their security.

The Marine Corps Combined Action Program (CAP) was created in 1965 with the hope that Marines with prior combat experience in Vietnam and the proclivity to work side by side with the Vietnamese would reside in villages, protect them from the Viet Cong, and train locals to protect their own villages. At the outset of each Marine's tour, he was trained in the political structure and culture of Vietnam but not in the language, which prevented the program from achieving maximum success. The fact that the CAP units were scattered also prevented the program from having maximum effect because they were not nearly networked enough to come to the aid of CAP units in other villages. Overall, the program had mixed success and was effective at attacking the Viet Cong Infrastructure (VCI) using indigenous forces. By the time the program was discontinued, the CAP, which represented only 1.5 percent of the Marines in Vietnam, accounted for 7.6 percent of enemy killed.[29]

The most successful and innovative approach to pacification was the CORDS program, which began in 1967 under the direction of Robert W. Komer, the Special Assistant to the President and Deputy for Pacification in Vietnam. The program's purpose was to unify all pacification efforts under one agency so as to eliminate the impasse between civilian and military bureaucracies, which it more or less accomplished. The success of the CORDS program relative to prior pacification efforts can be attributed to bureaucratic politics and the realization that a two-pronged approach was needed for pacification. Both civilian and military personnel brought strengths to CORDS that ensured its success—the military with its vast resources and manpower and the civilians with their understanding of the political nature of the war. Komer's assertive personality and his good rapport with the new MACV commander, General

---

[28] Blaufarb (1977, p. 106).

[29] Cassidy (2004, p. 77).

Creighton Abrams, contributed to the flexibility of the program, which allowed it to be more innovative and adaptive.

The timing for such innovation, although late in the conflict, was well timed for the Tet Offensive. Tet was a political victory for the insurgents, but their military losses were considerable and they required time to recuperate, consequently compromising their ability to hinder pacification. Recognizing this advantage, General Abrams implemented a clear-and-hold strategy, in which counterinsurgents would clear contiguous areas of insurgents and pacify the residents of that area.[30] By 1970, at least 90 percent of the population of South Vietnam was pacified, releasing ARVN to fight the insurgents and increasing civilians' incentives to cooperate with the counterinsurgents.[31]

The Phoenix Program was created in 1968 and was designed to sever the link between the ranking Viet Cong and the population by neutralizing or killing the VCI, which was engaged in recruiting and training of cadres.[32] Since the VCI was required to remain in contact with the population to maintain the traction of the cause, it was extremely visible and therefore vulnerable, unlike insurgents who could remain hidden whenever they felt the need to do so. The program was decentralized, with a South Vietnamese chief who was superior to his U.S. adviser, whose role was restricted to advising and calling in U.S. military support. Local indigenous forces were used to attack the VCI because they were familiar with the terrain and the inhabitants of the villages, which allowed the military to focus on the provision of security.

Throughout the program's existence, almost 82,000 VCI members were neutralized, a blow that a North Vietnamese general later described as "extremely destructive," although much of this can be attributed to the losses suffered by the insurgents during Tet.[33] Had the counterinsurgents understood the importance of the VCI in the early years of the insurgency when it was most vulnerable, they may have encountered greater success in their political and military operations.

During this pacification period, in an attempt to co-opt one of the insurgent causes, President Nguyen Van Thieu undertook such reforms as the election of local governments and the Land to the Tiller Act (1970) to ensure that the South Vietnamese had a political and economic stake in the future of the country. The Land to the Tiller Act was one of the most successful reforms carried out by the South Vietnamese government because it sought to eliminate one of the underlying causes of the insurgency and contributed to the amelioration of the economic situations of those it affected. In the first three years of the program alone, the percentage of South Vietnamese living or working on land they did not own dropped from 60 percent to 10 percent.[34]

---

[30] Andrade and Willbanks (2006, pp. 9–23).

[31] Boot (2002, p. 311); Andrade and Willbanks (2006, p. 22).

[32] Andrade and Willbanks (2006, p. 21).

[33] Boot (2002, p. 310); Andrade and Willbanks (2006, p. 21).

[34] Hunt (1995, p. 264).

Conclusions

The fall of South Vietnam in 1975 marked the end of the 30-year Vietnam War for independence and the first defeat for the United States. At the end of the conflict, 2–4 million South Vietnamese civilians were dead, as well as 58,000 Americans, 230,000 ARVN, and 550,000 North and South Vietnamese insurgents. The end of the war was a watershed for the United States not only because it created an aversion to fighting small wars and prevented the lessons of Vietnam from being institutionalized, but also because the United States was perceived as a paper tiger and a world power in retreat. After the fall of Saigon, Laos and Cambodia were quick to fall to communism and the Soviet Union became emboldened in other parts of the world.

That the insurgents must have been successful in their techniques because they won the war is a conclusion that should be avoided because many of their strategies and techniques met with mixed success. What is most important, however, is that they learned from their successes and failures and were extremely adaptable in their techniques. The insurgents were adept at combining the military and political aspects of their war, in classic insurgent fashion, and were able to successfully mobilize the people of South Vietnam as weapons against the government. Their decentralized organization provided social services where the government could or would not, which increased their legitimacy with those they sought to convert to their cause. Knowing that the United States had an advantage in conventional warfare, they lured the counterinsurgents away from population centers because they understood that the war was not over territory but over the loyalty and control of the population. And, finally, they used terrorism to coerce those who were undecided to join the cause, to intimidate those who were already committed, and to ensure that all cadres were thoroughly and continually indoctrinated with the cause. That said, the war was not won by the insurgents but by a conventional invasion from North Vietnam, as the last three years of the war were mostly fought by North Vietnamese conventional forces.

Likewise, one should not assume that the counterinsurgents lost because they were unsuccessful in their techniques, as pacification had mixed outcomes. The counterinsurgents eventually understood that they needed a unified, political-military solution for the war, but the time lost between the beginning of the insurgency and the implementation of these reforms allowed the insurgent infrastructure to mature and grow stronger. Once the importance of the insurgent infrastructure was established, the counterinsurgents were able to counter it by implementing programs to attack and gather intelligence from it and legally process detainees in a manner that minimized population alienation.

By not fully understanding the political cause behind the insurgency, the counterinsurgents were thus unable to successfully co-opt the insurgent cause until much later in the conflict. The counterinsurgents underestimated the effect that land reform, if carried out correctly in the insurgency's early stages, would have had on the insurgent cause and the effect that U.S. combat presence had on the erosion of the GVN's legitimacy and authority. The inability and sometimes unwillingness of the GVN to combat the insurgency and provide security and social services for the population and U.S. government support for these successive regimes further eroded what little authority they had.

Partly because of the dire state of the war when U.S. combat forces entered it in 1965, technology and firepower dictated the way the war was fought and inhibited the counterinsurgents from understanding that they needed to become more primitive in their techniques to combat the insurgency. Heavy reliance on technology contributed to the alienation of the population and hampered the counterinsurgents' ability to gather intelligence about the Viet Cong and VCI. However, not using the maximum amount of the most sophisticated technology would have been antithetical to the U.S. military culture and would have increased casualties and decreased public support. Because of this preference for conventional warfare on the heels of U.S. involvement in two large conventional wars in the 1940s and 1950s, many potentially successful counterinsurgency programs were dismantled and their resources redirected to purely offensive military operations. Additionally, the U.S. view that the war in Vietnam was a sideshow for the larger Cold War battles had leaders rotating troops frequently to give them battlefield experience, rather than leaving them in place long enough to learn about their environments.

On the other hand, the counterinsurgents were eventually willing and able to use such insurgent tactics as small-unit patrols and the formation of mass associations, which demonstrated their eventual open-mindedness to new ideas and their ability to take advantage of enemy setbacks, as in the post-Tet era. Ultimately, the unity of command for CORDS and the advantage of its timing during a period of severe enemy weakness allowed the counterinsurgents to make inroads in pacification, while the use of indigenous forces and locals freed the military up for security operations and added legitimacy to South Vietnamese counterinsurgent operations.

Had the counterinsurgents been able to enact the same reforms in the early 1960s that they did later that decade, they may have encountered greater success and their lessons may have been preserved for future generations engaged in counterinsurgency operations. Unfortunately, because of the outcome of the war, this was not the case. One may conclude that some of the lessons that can be drawn from counterinsurgent operations in Vietnam include the recognition of the marriage between the political and military in an insurgency, the desirability of a long-term perspective open to many possibilities, harsh objectivity in the face of failure, and a penchant for adaptability, as well as the fact that a successful pacification campaign run by foreigners is not a viable alternative to pacification and maintenance of a political base by the host nation.

CHAPTER FIVE

El Salvador (1980–1992)

Angel Rabasa[1]

Origins and Characteristics of the Insurgency

El Salvador has a long history of social violence. The country is the most densely populated mainland country in the Western hemisphere.[2] Demographic pressures and restricted access to land generated a great deal of social unrest. In 1932, a communist-inspired uprising claimed at least 10,000 lives.[3] Since the overthrow in 1944 of General Maximiliano Hernández Martínez, the last of the traditional Salvadoran military dictators, the armed forces as an institution became the predominant political actor in El Salvador. The traditional landed elite—the so-called "oligarchy"—had not exercised power directly since the 1920s, although it continued to have an important voice in economic and social policy. After a period of reform in the 1960s, the military established a political party, the National Conciliation Party (PCN) modeled after the Mexican ruling party, the Revolutionary Institutional Party (PRI), which managed the political process and attempted to mediate the interests of different social sectors. (See Figure 5.1 for a map of El Salvador.)

This system was successful in maintaining stability from the late 1940s until the early 1970s, notwithstanding El Salvador's demographic problem, highly inequitable land tenure system, and high levels of underemployment. The system began to show signs of strain in the 1970s, when new parties and organizations demanded an opening of the political system. The Christian Democratic leader José Napoleón Duarte won the 1972 presidential election at the head of a center-left coalition, but the government manipulated the election returns to ensure the victory of the official candidate.

---

[1] Unless otherwise noted, the political and military assessments in this paper are based on the author's observations as Regional Security Officer for Latin America in the Bureau of Political-Military Affairs, U.S. Department of State, during the first Reagan administration.

[2] It has 277.6 inhabitants per square kilometer, more than Haiti (246.0), over five times the population density of Honduras (52.4), and seven times that of Nicaragua (38.1). Some small Caribbean islands have higher population densities ("Population Density: Latin America," n.d.).

[3] The 1932 uprising was an important event in the leftist narrative in El Salvador. The main guerrilla coalition in the 1980s, the Farabundo Martí Nacional Liberation Front (FMLN) took its name from the communist leader killed after the uprising.

Figure 5.1
El Salvador

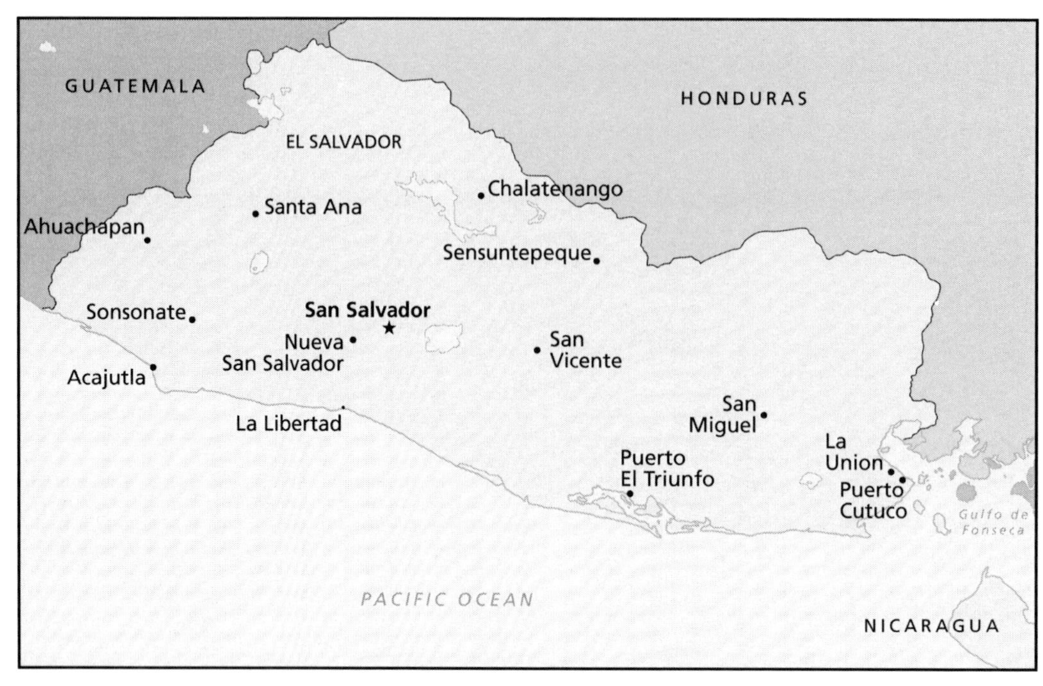

RAND *OP185-5.1*

The 1970s also saw the emergence of Marxist armed organizations. A split in the Communist Party of El Salvador (PCES) led to the establishment of the Popular Liberation Forces (FPL), led by Salvador Cayetano Carpio, a former PCES secretary-general. Carpio's strategy was Vietnamese-style people's war. In line with this strategy, the FPL began to carry out "armed propaganda" (i.e., kidnappings and assassinations) and to establish a base in the mountains of Chalatenango department on the Honduran border. A second guerrilla group was the Cuban-line People's Revolutionary Army (ERP), which derived inspiration from Guevarist theories of guerrilla warfare. The ERP followed the developed ties with several guerrilla organizations in Latin America, including the Tupamaros of Uruguay, the People's Revolutionary Army of Argentina, and the Movement of the Revolutionary Left (MIR) of Chile. A split within the ERP in 1975 resulted in the formation of another group, the Armed Forces of National Resistance (FARN).[4] The universe of armed Marxist groups that proliferated at this time also included the small Trotskyite Revolutionary Party of Central American Workers (PRTC).

Throughout the 1970s, the activities of these groups could be considered political-criminal but did not amount to an insurgency. The groups financed themselves through kidnappings for ransom of Salvadoran and foreign businessmen. The most notorious of these actions was the FPL's kidnapping and murder of Foreign Minister Mauricio Borgonovo in

---

[4] Kruger (1981).

May 1977.[5] They all established political front organizations to carry out propaganda activities and recruitment. The ERP established the LP-28 (Popular Leagues February 28).[6] The LP-28 undertook an important role in subverting the human rights movement in El Salvador by promoting an ERP member into a position of leadership within the Salvadoran Commission of Human Rights (CDHES). The FARN took over the FAPU (Unified Popular Action Front), which had been formed in 1974 by the ERP and two radical Catholic priests. The FAPU conducted numerous strikes, marches, and propaganda distribution projects in the labor movement. The Moscow-oriented PCES remained active in El Salvador's political system through its recognized legal front, the National Democratic Union. The Communist Party itself was illegal.[7]

There was, therefore, a substantial extreme left infrastructure in place in El Salvador when, in October 1979—three months after the overthrow of the Somoza regime in Nicaragua—a group of young officers toppled the government of General Carlos Humberto Romero. The officers appointed a center-left junta and announced a program of political and economic reforms. The Sandinista victory in Nicaragua, however, followed by the fall of Romero, encouraged the Salvadoran Marxists in their belief that a revolutionary tide was under way in Central America that, as the official history of the FMLN states, "opened the possibility of resolving the problem of [seizing] power."[8] The response of the radicals was to step up their actions to bring about the collapse of the new government.

The revolutionary strategy pursued by Cuba in Nicaragua in 1978–1979 was also implemented in El Salvador. First, the traditionally splintered Marxist groups were required to unite as a condition of Cuban support. In May 1980, the leaders of the different factions met in Havana and formed a political-military command, the Unified Revolutionary Directorate (DRU), as the central executive arm for political and military planning. The DRU established its headquarters near Managua, Nicaragua, and helped to direct planning and operations and coordinate logistical support for its forces in El Salvador.

A further step toward unification was taken later in the year when the FPL, ERP, PCES, FARN, and PRTC formed the Farabundo Martí National Liberation Front. The unification of these groups was a prelude to the launching of a so-called "final offensive" in January 1981.[9] The military campaign was supported by a political front, the Democratic Revolutionary Front (FDR), a "broad coalition" led by the extreme left but including elements of the non-communist left. The popular front tactic helped the guerrillas to pose as a broadly based move-

---

[5] Other victims included former Provisional President Osmin Aguirre, who had participated in the suppression of the 1932 uprising, the rector of the University of El Salvador, a former president of the legislative assembly, and the South African ambassador.

[6] The LP-28 took its name from the date, February 28, 1977, when a number of demonstrators were killed in a protest against electoral fraud in the election of President Carlos Humberto Romero.

[7] Kruger (1981); Gettleman et al. (1981).

[8] Frente Farabundo Martí (n.d.).

[9] See General Order No. 1, issued by FMLN Commander Salvador Cayetano Carpio on January 11, 1981, in "The Final Offensive," in Gettleman et al. (1981, pp. 118–120). The FMLN called for a general strike in cooperation with the FMLN fighters.

ment and to attempt to isolate the Salvadoran government and deprive it of international support.[10]

The Salvadoran insurgency began, then, based on a political-military strategy that paralleled the successful Sandinista-led insurgency in Nicaragua. (The Nicaraguan insurgency was a broad-based movement that incorporated elements on the political spectrum from conservatives to Marxists, but the Sandinistas controlled the armed component.) Unlike the situation in Nicaragua in 1979, however, there was no broad-based support for armed insurgency in El Salvador. Despite substantial weapons deliveries from Cuba and Nicaragua and the lack of readiness of the Salvadoran military, which was then under a U.S. arms embargo, FMLN "final offensives" in January 1981 and again in early 1982 and in 1989 failed to overwhelm the Salvadoran government. As the Salvadoran armed forces grew more capable with the influx of U.S. military assistance after 1981, the FMLN reverted to classical guerrilla tactics and the war developed into a stalemate.[11]

Strengths of the Guerrillas

The FMLN carried out a robust insurgency for a period of eight years, despite significant U.S. military support for the Salvadoran government. The insurgency was ended by a peace agreement that gave the FMLN a role in the political life of the country. Some of the factors that account for the insurgency's resilience are the following:

1. *Ability to generate significant forces.* By 1983, the FMLN was able to field a force of 9,000 to 12,000 combatants (giving the Salvadoran Army less than a 4:1 overall advantage) and could mount battalion-level operations involving as many as 500 to 900 fighters. Despite improvements in the Salvadoran armed forces as the result of the U.S. train-and-equip program, the guerrillas maintained the ability to mass forces, overwhelm isolated garrisons, and ambush relief columns. To coordinate operations, the FMLN established a sophisticated internal communications system linking its fronts throughout the country.[12]
2. *Adaptability.* After U.S. assistance to the Salvadoran military began to turn the military balance against the guerrillas, the FMLN adapted to the growing strength and capabilities of the Salvadoran armed forces by breaking down large columns into smaller squads and adopted new tactics, including increased use of land mines and homemade weapons, attacks on U.S. military personnel and Salvadoran military training centers, and urban operations. The FMLN developed special forces units, known collectively as

[10] *The Report of the President's National Bipartisan Commission on Central America* (1984, p. 107); Frente Farabundo Martí (n.d.).

[11] Unless otherwise noted, the political and military assessments in this paper are based on the author's observations as Regional Security Officer for Latin America in the Bureau of Political-Military Affairs, U.S. Department of State, during the first Reagan administration.

[12] U.S. Library of Congress (1988a).

Fuerzas Especiales Selectas—FES (Select Special Forces), that were responsible for some of the most spectacular attacks of the war, such as the raid on the Salvadoran Air Force main base at Ilopango in January 1992, in which FMLN sappers destroyed much of the Salvadoran Air Force on the ground.[13]

3. *External support.* The arms pipeline from Cuba and Nicaragua was critical in sustaining the Salvadoran insurgency. Before the "general offensive" of January 1981, tons of modern weapons, primarily U.S.-made arms from captured stockpiles in Vietnam, were delivered covertly to guerrilla forces in El Salvador.[14] Arms shipments continued throughout the period of the armed conflict, usually transported by small boats and planes across the narrow Gulf of Fonseca, shared by El Salvador with Honduras and Nicaragua.
4. *Use of sanctuaries.* Throughout the war, the rebels operated from bases in the mountainous areas bordering Honduras (including refugee camps in Honduras), in the rugged Guazapa and San Vicente volcanoes, and along the seacoast in Usulután department, where a multitude of inlets facilitated the smuggling of arms and supplies from Nicaragua. Large-scale government "hammer and anvil" operations failed to clear these areas on a sustained basis.

Weaknesses of the Guerrillas

1. *Lack of mass support.* The FMLN and its political front, the FDR, never enjoyed majority support among the population. Lack of mass support was manifested in the failure of several "final offensives" (1981, 1982, and 1989) to provoke a mass uprising as anticipated by the FMLN. The guerrillas' strategy of systematic attacks on the country's infrastructure also cost them a loss of popular support. The massive turnout at the 1982 and 1984 elections, despite the threat of violence, could be considered a repudiation of the guerrillas' methods as well as their aims.
2. *Lack of political legitimacy.* The guerrillas were weakened politically by a sustained democratic process that brought to power a government with viable democratic credentials. This process began with elections for a constitutional assembly in 1982 and presidential elections in 1984 that brought to power the Christian Democratic Party and its leader, José Napoleón Duarte. (The FMLN demobilized as a result of the 1992 Peace Accord and participated in the 1994 elections and in subsequent elections.)
3. *Disunity.* Although the different guerrilla forces were formally unified in the FMLN, each retained its separate identity, doctrine, and area of operations. Moreover, suspicions and rivalries among the groups' leaders continued and sometimes turned deadly.

---

[13] The attack destroyed six of the air force's 14 UH-1H helicopters and eight fixed-wing aircraft. For a detailed analysis of FMLN strategy and tactics, see Bracamonte and Spencer (1995).

[14] U.S. Library of Congress (1988a).

The deep divisions within the FMLN were revealed by the murder of Mélida Amaya Montes (Comandante Ana María), the FPL's second in command, and the suicide of FPL leader Carpio in Managua in April 1983. (Montes and Carpio had clashed over strategy.)[15]

4. *Inability to prevent U.S. assistance to the Salvadoran government.* From the beginning, the guerrillas and their supporters were aware that U.S. support over time could tilt the military balance in favor of the Salvadoran government (as eventually happened). Therefore, the guerrilla strategy in early 1981 was to overwhelm the government before U.S. assistance to El Salvador could resume. (U.S. military assistance to El Salvador was resumed by the Carter administration on January 16, 1981.)[16] Until the signing of the Peace Agreement in 1992, FMLN supporters and sympathizers mounted a campaign in the United States and Europe to bring about the end of U.S. and international support of the Salvadoran government—a campaign facilitated by death squad activities and high-profile incidents such as the 1980 assassination of Archbishop Oscar Romero and the 1989 murder of six Jesuit priests.[17]
5. *Reliance on outside support.* What was one of the strengths of the insurgency—Cuban and Nicaraguan support—became a liability when that support declined as a result of changes in the international environment after 1989. The retrenchment and subsequent collapse of the Soviet Union and the end of Soviet assistance to Cuba and Nicaragua raised the risks and costs to those countries of continuing to support the Salvadoran insurgency. The Sandinista government of Nicaragua also came under additional pressure because of U.S. support of the Nicaraguan resistance (which some U.S. officials presented as reciprocal treatment for the Sandinistas for their support of the Salvadoran insurgency).

Strengths of the Government

The strengths and weaknesses of the government side were, in some respects, the obverse of the guerrillas' strengths and weaknesses. The main strengths of the Salvadoran government during the civil conflict, which in the end enabled it to prevail, were the following:

1. *Political legitimacy.* Before the 1979 coup, El Salvador—like most Central American countries except Costa Rica—had no tradition of genuine democratic governance. The country was run by a series of military governments, sometimes overtly but often with a façade of constitutional government and (manipulated) elections. As noted above, in the 1980s, El Salvador began a democratic experiment with a series of free elections (constitutional assembly and legislative elections in 1982, 1985, 1988, and 1991; and

[15] On December 27, 1980, Comandante Fermán Cienfuegos of the FARN announced that a final offensive would be launched before President Ronald Reagan's inauguration on January 20, 1981 (United States Institute of Peace, 1993).

[16] National Security Archives (1989, p. 38).

[17] In 1991, two military officers were convicted of the murder of the Jesuits.

presidential elections in 1984 and 1989) that delivered a government that had legitimacy and broad popular support.

2. *Duarte's leadership.* The outcome of the civil conflict in El Salvador—in which the insurgency failed to prevail—would have been much more problematic without the leadership of President Duarte. As leader of the Christian Democratic opposition to the pre-1979 military regime, Duarte had great personal credibility. His leadership was instrumental in keeping El Salvador on a democratic track, curbing excesses by the security forces, and maintaining international support for El Salvador.
3. *International support.* U.S. and international support was critical to the survival of the Salvadoran government. Despite the international campaign against international support for the Salvadoran government discussed above, U.S. aid to El Salvador rose from $264.2 million in fiscal year (FY) 1982 to an estimated $557.8 million in FY 1987. The Federal Republic of Germany also provided economic and military assistance to El Salvador.
4. *Restructuring of Salvadoran military and security forces.* At the onset of the insurgency, the Salvadoran armed forces were a barracks-bound, defensively minded organization with severe deficiencies in command and control, tactical intelligence, tactical mobility, and logistics. The U.S. security assistance program was designed to address these deficiencies and transform the Salvadoran armed forces' strategy, doctrine, training, and equipment, with a minimum of direct U.S. involvement. (The ceiling on the number of U.S. military trainers in El Salvador was fixed at 55 personnel throughout the armed conflict.) For most of the U.S. military assistance program to El Salvador, the training mission was a Special Forces mission; it represented the U.S. government's awareness that the United States was not ready to send conventional forces so soon after the end of the Vietnam War. The brigade Operational Planning and Assistance Training Team (OPATT) mission lasted 12 years and was one of the longest-running U.S. Special Forces missions. The Special Forces involvement had much to do with the success of the U.S. security assistance program to El Salvador.[18]

In the context of this military assistance program, the United States trained the Salvadoran Army's "rapid reaction" battalions that played a key role in moving the Salvadoran armed forces out of the strategic defensive.[19] In formerly guerrilla-held areas, the government implemented a civic action program that consisted of rebuilding the social and economic infrastructure and training civil defense units to protect key targets and free the military to engage in offensive operations.[20] Finally, after the counterinsurgency effort gained momentum in the mid-1980s, the morale of the government forces was usually high and not on the verge of collapse, as alleged by some critics of the war.[21]

---

[18] Comments by Robert Everson, review of manuscript, January 2007. See also Bailey (2004).

[19] U.S. Library of Congress (1988b).

[20] Maitre (1987).

[21] See, for instance, Bracamonte and Spencer (1995).

Weaknesses of the Government

Although the government's forces were able to hold their own in a decade-long armed conflict, they evidenced significant weaknesses that made the outcome at times uncertain.

1. *Strategic vulnerabilities.* These vulnerabilities derived from El Salvador's geographic and demographic configuration. Unlike, for instance, Colombia, where FARC formations generally operate in peripheral areas at considerable distances from major population centers, in El Salvador, the guerrilla strongholds were within striking distance of the country's main cities and critical infrastructure. The eastern region of El Salvador, the FMLN's main area of operations, suffered the brunt of the attacks, but no part of the country was immune. In its 1989 general offensive, for instance, the FMLN attacked military installations in major cities throughout the country, including the capital, San Salvador.
2. *Lack of border control.* These strategic vulnerabilities included lack of control of El Salvador's border with Honduras and the maritime border with Nicaragua in the Gulf of Fonseca. Salvadoran refugee camps on the Honduran side of the border were used as sanctuaries by FMLN guerrillas.[22]
3. *Overreliance on air mobility.* Because of the large-scale guerrilla attacks on bridges and roads, especially in the initial stages of the conflict, the Salvadoran military depended heavily on air transport. This was both a strength and a weakness. The heavy use of UH-1H helicopters for mobility resulted in a high down ratio—particularly since many of the helicopters delivered by the United States to El Salvador were aging and frequently of poor quality. Similarly, most of the A-37s required excessive, unscheduled maintenance time.[23] Fortunately for the government side, the Salvadoran rebels never introduced surface-to-air missiles, which would have compromised the use of aircraft in transport and close air support roles by government forces.
4. *Imperfect government control over security forces and continued human rights violations.* Death squad activity, some of it allegedly linked to elements of the security forces, continued throughout the period of the insurgency (although the number of political murders declined significantly after Duarte became president in 1984). Allegations of human rights violations constituted a significant obstacle to the efforts of the Reagan and first Bush administrations' efforts to maintain congressional and public support for the government of El Salvador.
5. *Uncertainty of continued U.S. assistance.* Opposition among sectors of the U.S. Congress to continued assistance to El Salvador created uncertainty about the future of U.S. assistance. The Salvadoran armed forces responded to the uncertainty of continued funding by hoarding essential materials, to the detriment of warfighting.[24]

---

[22] Some of these camps were in the bolsones ("pockets")—small, demilitarized strips of land along the border disputed by El Salvador and Honduras.

[23] Maitre (1987, pp. 130–131).

[24] Coates (1991).

6. *Economic crisis.* The conflict (aggravated by FMLN attacks on the economic infrastructure) depressed agricultural and industrial production. Despite a program of economic stabilization, the economic crisis deepened in the mid- to late 1980s. Unemployment increased from single digits to 33 percent of the labor force between 1978 and 1985, and real wages declined by about one-third between 1983 and 1987.[25] Economic distress led to the Christian Democrats' loss of their parliamentary majority in 1988 to the victory of the right-wing ARENA party in the 1989 presidential election.

Conclusions

El Salvador was the theater of the most successful U.S.-supported counterinsurgency effort since the 1960s.[26] The insurgency was defeated despite significant weaknesses on the government side—which included poorly trained and equipped government forces at the beginning of the insurgency, an almost total lack of civil defense (a key element in the successful Guatemalan counterinsurgency campaign), and significant external support for the insurgents. This external support, as noted above, enabled the FMLN to generate a large guerrilla force (by Salvadoran standards) that, as late as 1983, operated at will throughout the countryside and inflicted heavy casualties on government forces.[27] What turned the tide was a political-military strategy implemented by the U.S. and Salvadoran governments.

On the political side, the Salvadoran government was able to assemble a broad-based coalition from conservatives to Christian Democrats that was led for much of the war by José Napoleón Duarte, first as a member of the junta that replaced the Romero government and then as president of El Salvador (1984–1989). Duarte's credibility as a democratic reformer was indispensable in maintaining U.S. and international support for El Salvador during the war.[28] On the military side, the United States implemented a train-and-equip program that transformed the Salvadoran armed forces from a static, defensively minded organization into a force capable of offensive operations against the guerrillas. The transformation of the Salvadoran armed forces involved the creation of several 1,000-man rapid-reaction battalions trained at Fort Benning, Georgia, and at the Regional Military Training Center (RMTC) in Puerto Cas-

---

[25] U.S. Library of Congress (1988b).

[26] Other notable U.S.-supported counterinsurgency successes since World War II were the Greek Civil War (1946–1949), the Huk insurgency in the Philippines (1946–1955), the Venezuelan insurgency (1960–1968), and Ernesto "Che" Guevara's ill-fated foray into Bolivia (1966–1967). The urban terrorist campaigns waged by the Montoneros in Argentina and the Tupamaros in Uruguay and the Shining Path insurgency in Peru were put down by local forces without significant U.S. involvement.

[27] Major government defeats at the end of 1983 included the destruction of the Fonseca and Tecana battalions, about 280 strong each, by much larger guerrilla forces; the capture of the Fourth Infantry Brigade headquarters at El Paraiso, Chalatenango; and the destruction of the Cuscatlan bridge on the Pan-American Highway, a vital link in the transportation infrastructure. The defeats were symptomatic of coordination and other problems in the Salvadoran Army at the time (Maitre, 1987).

[28] Duarte played this role at great personal cost. The guerrillas kidnapped his daughter in 1985. He died of cancer in 1990.

tilla, Honduras. (By late 1983, the United States had trained 900 Salvadoran officers, or half the entire officer corps.)[29]

The key strategic problem was interdicting Sandinista support for the Salvadoran guerrillas. In the final analysis, there may not have been any permanent solution to the Salvadoran Army's strategic problem short of striking at the Nicaraguan root of the guerrilla movement's military strength—a task beyond the capabilities of the Salvadorans alone. This conundrum was resolved by the emergence of the U.S.-supported armed resistance movement against the Sandinistas—the so-called "contras." The development of an anti-Sandinista insurgency turned the tables on the Sandinista government, which was forced to fight for its own survival while supporting the Salvadoran insurgency. Finally, the geopolitical change brought by the collapse of the Soviet Union and the realization, on the part of the Salvadoran guerrillas, that a military victory was increasingly out of reach created the conditions for a negotiated end to the conflict.

---

[29] U.S. Library of Congress (1988b).

CHAPTER SIX

Jammu and Kashmir (1947–Present)

Paraag Shukla

Origins and Characteristics of the Insurgency

Sovereignty over the Indian state of Jammu and Kashmir (J&K) has been disputed ever since India and Pakistan gained their independence in August 1947. As laid out by the plan for partition under the Indian Independence Act of 1947, rulers of the princely states were allowed to choose either to stay within India or move to Pakistan. (See Figure 6.1 for a map of Jammu and Kashmir.)

On October 22, 1947, armed tribesmen and troops from Pakistan's North-West Frontier Province crossed the border into Kashmir, aiming to capture Srinagar, the capital of J&K. Unable to deal with this invasion, Maharaja Hari Singh formally signed an Instrument of Accession on October 26, 1947, making Jammu and Kashmir a state of India.

War ensued between India and Pakistan, continuing for over a year until the UN Security Council established the UN Commission for India and Pakistan (UNCIP) and became involved in the conflict. A cease-fire was arranged by the UN for December 31, 1948, and the UNCIP Resolution of August 13, 1948, was formally adopted by the following January. When the UN ordered both sides to hold at their current positions, Pakistani forces had not yet completed their withdrawal from the territory they had seized by force. As a result, they were able to acquire over one-third of Kashmir.

Since that time, India and Pakistan have fought two additional declared wars, in 1965 and 1971.[1] Following the 1971 war, the leaders of both countries signed the Simla Agreement, stipulating that they would not attempt to alter the ad hoc, newly dubbed line of control (LoC).

A movement for independence also exists within J&K and has steadily grown since the late 1980s. After the failed Soviet campaign in Afghanistan, a large number of mujahideen shifted east to J&K with a great deal of support from the Pakistani government. Furthermore, with the continual encouragement and support of the Pakistani Directorate for Inter-Services Intelligence (ISI), the mujahideen began to view J&K through a religious lens as a jihad. In the 1990s, tensions in the region increased as did the level of violence. Coordinated attacks

[1] Although the 1971 war was initiated by genocide in East Pakistan, the theater of fighting rapidly expanded to include J&K.

Figure 6.1
Jammu and Kashmir

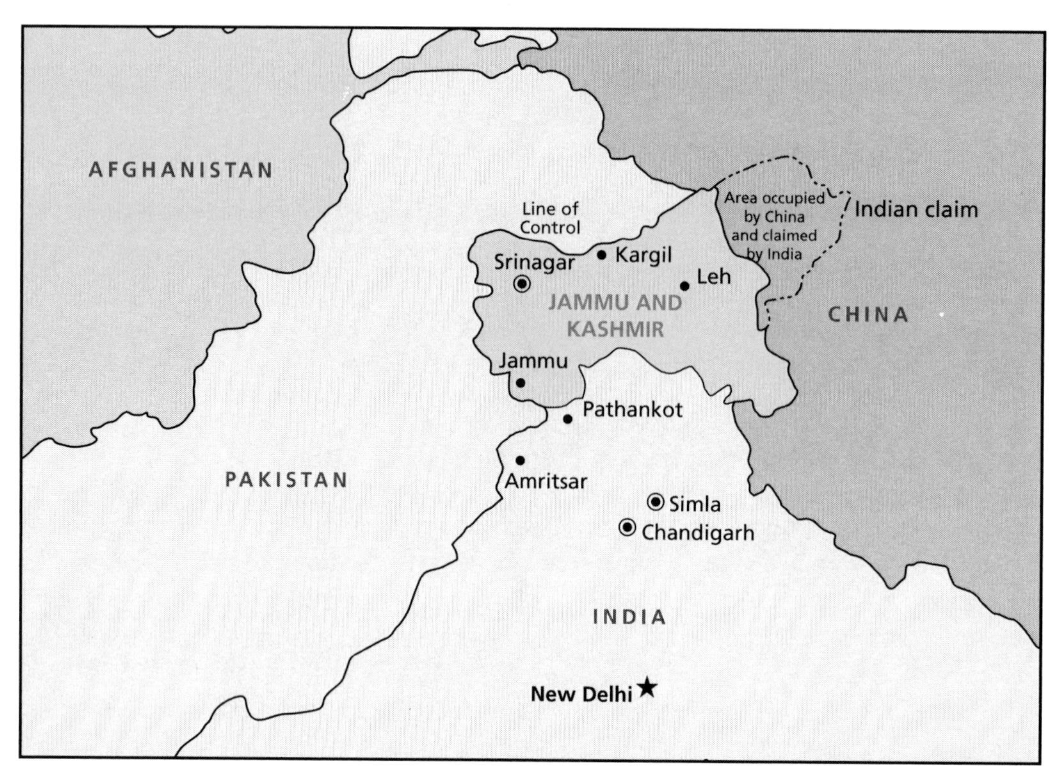

RAND *OP185-6.1*

were carried out on non-Muslims in the Kashmir Valley and, as a result, Indian security forces immediately stepped up their presence. Indian paramilitary forces found themselves unable to fully deal with the wide-scale militancy and so the Indian Army was redeployed to the Valley to commence COIN operations.

In 1998, both countries openly tested nuclear weapons, giving rise to worries in the international community about the addition of nuclear weapons to an already unstable region. Soon after, both New Delhi and Islamabad took open strides toward forming peaceful relations. Indian Prime Minister Atal Bihari Vajpayee accepted the invitation of his counterpart in Pakistan, Nawaz Sharif, and traveled to Lahore to continue peace talks.

Optimistic hopes for peace were shattered a few months later, when Pakistani soldiers infiltrated across the line of control in the Kargil area of J&K and seized Indian peaks along the Srinagar-Leh Highway, resulting in a fierce and bitterly fought limited war.[2] One of the

---

[2] Indian Intelligence determined that one main thrust of the Pakistani plan in Kargil was to seize as much territory as possible and then hold off Indian forces until the UN intervened. Islamabad believed that both nations' nuclear status would accelerate the involvement of the international community. The UN would presumably follow precedent and order both sides to hold their positions, an action that would be taken almost certainly before the Indian forces could finish the lengthy process of flushing out the intruders. Much like the outcome of 1947–1948, Pakistan would therefore be able to attain more vitally important Indian territory.

chief masterminds of the Kargil incursion, Pakistani General Pervez Musharraf, overthrew Sharif in a bloodless military coup and seized power.

A terrorist attack on the state legislature in J&K in October 2001 and another attack on the Indian parliament in December brought tensions to an unprecedented high. The Indian military immediately redeployed the vast majority of its forces to its border with Pakistan and both nations braced for the possibility of another war. Amid the already heightened tensions, another terrorist attack, this time targeted at the wives and children of Indian soldiers, occurred in Kashmir on May 14, 2002. Pakistan insinuated the possible use of nuclear weapons after India issued stern warnings for Islamabad to discontinue support of the terrorist elements involved. Although the situation was gradually deescalated, the future of the region remains unclear.

Strengths of the Insurgents

Historically, Islamic militant groups and paramilitary forces have directly helped the Pakistani government obtain strategic and military objectives.[3] In turn, Islamabad has granted such organizations very wide latitude to operate throughout the country.

Pakistan's regional proximity to Afghanistan and the Middle East has come under the spotlight since the terrorist attacks on September 11, 2001. President Musharraf, although a U.S.-declared ally in the Global War on Terror, has continued to maintain an ambiguous stance on terrorist activities and cross-border infiltration into J&K.

In 1999, when many Indian battalions were rapidly redeployed to the Kargil area to battle the cross-border intrusion, there was an increased flood of militants into the Kashmir Valley. As a result, the successful outcome of the Kargil conflict was underscored by disappointing setbacks in COIN operations across J&K.

There were many attempts by the ISI to disrupt the 2002 elections by finalizing the formation of the Kashmir Liberation Army, an organization that would establish a unified command structure and communication network for terrorist groups across J&K. The existence of such an organized and well-established terrorist network in the region would undoubtedly reduce the visible link to the ISI's direct involvement in cross-border terrorism and therefore mask any semblance of guilt.[4]

Following the decline of the Jammu and Kashmir Liberation Front (JKLF), there remain over a dozen terrorist groups currently reported to be active in J&K. The percentage of foreign contingents among the ranks of these outfits has risen sharply, from 6 percent in 1992 to over 50 percent. Three such groups—*Lashkar-e-Toiba* (LeT—"Army of the Pure"), *Jaish-e-Mohammad* (JeM—"Army of the Prophet Mohammad"), and *Harkat-ul-Mujahideen* (HuM—"Movement of Holy Warriors")—are directly supported by the Pakistani ISI in hopes of repro-

---

[3] Examples also include the direct implementation of militant forces alongside Pakistani regulars in the 1947–1948, 1965, and 1999 India-Pakistan wars.

[4] South Asia Terrorism Portal (2003).

ducing the successes of the Afghan insurgency against the Soviets.[5] These three groups have been consistently effective in carrying out their operations and are suspected to have formed links to Osama bin Laden's al Qaeda network.

Lashkar-e-Toiba has the reputation as being one of the largest and most brutal terrorist organizations in the state. None of its members, who number nearly 1,500, are of Kashmiri origin.[6] Terrorists from LeT assaulted the Cantonment in Delhi's Red Fort in 2000, and the group is suspected, along with members of JeM, to have carried out the attacks on the Indian parliament in Delhi on December 13, 2001.

The organization has two objectives, both driven by firm ideology. It wants to establish a fundamentalist theocracy and to effectively expand and export its local struggle to the entire country. In the LeT's view, the insurgency in Jammu and Kashmir is religious in nature.

In the fall of 2002, *Jaish-e-Mohammad* was classified by U.S. Secretary of Defense Donald Rumsfeld as one of the "deadliest organizations in the terrorist underworld."[7] Launched in 2000, the JeM is a relatively recent addition to the array of terrorist groups operating in Jammu and Kashmir and has been deemed responsible for the 2001 Indian parliament attacks mentioned above. This attack is the only instance in which the group has operated outside Jammu and Kashmir. The group has vowed to "liberate" Kashmir and other important religious sites across the country.[8]

Harkat-ul-Mujahideen took numerous tourists and security forces personnel hostage in the mid-1990s to try to compel the government to release its arrested leaders. For this same purpose, suspected HuM terrorists hijacked Indian Airlines Flight IC-814 in 1999. They diverted the plane and, with the support of the Taliban regime, flew to Kandahar, Afghanistan. Despite a decade of fairly consistent action, the HuM's operational abilities have been weakened since 2000 and the creation of the JeM (which has drawn many HuM recruits). Despite this, it still remains at the forefront of military activity in the state.

The initial Indian response to the militancy was heavy-handed and therefore served to initially alienate much of the Muslim population. Relaying promises of financial rewards and protection for families from the reported abuses of the Indian forces, insurgent groups were able to consistently recruit young men. In the early 1990s, when the ISI began to drape the veil of religious jihad over the local unrest, it also began considerable information operations in Jammu and Kashmir. Almost all terrorist groups, including the three mentioned above, are very active in recruiting from rural areas, where the presence of security forces is scarce or inconsistent.

---

[5] Two of these groups, LeT and JeM, were in fact created and raised by the ISI (Brennan et al., ongoing RAND Corporation research).

[6] Brennan et al. (ongoing research).

[7] South Asia Terrorism Portal (2003).

[8] South Asia Terrorism Portal (2003).

Weaknesses of the Insurgents

Although the insurgents across Jamma and Kashmir claim to be fighting on behalf of the civilian population, none has made efforts to provide any social services or security to villages in the Valley. Their sights have been set on both security forces and civilian targets. Violence against the population, regardless of religious orientation, is commonplace.

Terrorist organizations in Jammu and Kashmir have commonly targeted noncombatants: In 2002 alone, over 90 women and children were directly targeted and killed in an estimated 50 incidents; many more were wounded.[9]

To further instill fear and publicity and therefore coerce the locals' loyalty, terrorists have, in many cases, mutilated their victims. In 2002, terrorist organizations issued a decree for girls to not go to educational institutions and to remain veiled. To force compliance, terrorists beheaded three girls and threw acid at others who were not wearing veils.[10] Owners of stores selling alcohol have also been attacked following the issuance of similar decrees. Such incidents have helped further sap widespread support from the locals.

All three of the groups mentioned above have also employed the use of fedayeen (sacrifice) squads on multiple occasions. These missions are considered to be high risk rather than suicidal. In many instances, small groups of terrorists infiltrate an operational area of security forces, fortify themselves in a favorable position, and proceed to kill as many security personnel as possible before being cut down. Lieutenant General Arjun Ray, a retired officer who served for 38 years including in Jammu and Kashmir, recalls that "Kashmiri militants generally put up a fight when their group is stiffened by a few mercenaries. Left to themselves, they prefer to hit and run."[11]

Although the use of suicide bombers has not been widely implemented in J&K, the LeT has shown itself to be skilled in the use of improvised explosive devices (IEDs). The targets of these land mines and other explosives include military convoys and other vehicles belonging to security forces, as well as civilian targets.

Characteristics of the Counterinsurgent Forces

The Indian Army is divided into several command areas: Northern, Western, Eastern, Southern, Central, Army Training Command, and a newly constituted Andaman and Nicobar Joint Command. Each of these is commanded by a lieutenant general.

Northern Command has its headquarters in Udhampur, Jammu and Kashmir, and has been a frontline for each of India's wars since independence (including the 1962 Sino-India War). It has also bore the brunt of COIN operations and guarding the LoC against infiltration and, in 1999, was tasked with flushing out the Pakistani intrusions in Kargil.

[9] South Asia Terrorism Portal (2003).

[10] Brennan et al. (ongoing research).

[11] Ray (1997).

In 1994, when the militancy was at its peak, the Indian government approved the creation of a new military unit, the Rashtriya Rifles (RR), to deal with insurgency, the security of rear areas, and other special operations. Regular army units could therefore be released from the consistent attrition common during frequent COIN operations and follow their standard doctrine for deployments. RR personnel, although drawn from army ranks, were retrained in people-centric operations. In 2000, Delhi authorized the expansion of the RR by 30 battalions, planning to bring the total up to 66 after five years.

Weaknesses of the Counterinsurgents

By 1990, it became apparent that J&K was gripped by a higher level of insurgency than had been experienced before. It was quickly determined that the regular army, already engaged in curtailing cross-border infiltration along the LoC, could hardly cope with the full range of COIN operations. Furthermore, the heavy-handed tactics of a purely military approach to the insurgency was resulting in inadvertent civilian deaths and collateral damage. This, coupled with the consistent difficulty of determining the location of militants among the population, was causing increasing alienation and discontent.

Interservice quarrels in 1993 following the creation of a Unified Headquarters (UH) in J&K caused additional setbacks. Created to coordinate COIN operations among the army, paramilitary, and police forces, the UH was ineffective and counterproductive. As India's oldest paramilitary force, the Border Security Force (BSF) wanted to place the Rashtriya Rifles under its command. The RR units, staffed by regular army personnel, looked down on the BSF's abilities and dismissed such notions. Furthermore, early RR battalions lacked cohesion, as they were assembled by amalgamating soldiers from different battalions.[12] Northern Command also began the challenging task of reorienting soldiers to COIN operations by adapting a political approach that concentrated on the population rather than on the militants. Language differences, unit traditions, and a high degree of equipment variance at the battalion (or even the company) level were all barriers to successful early deployments in the Valley. It was not until after the Kargil War that Northern Command was able to act on lessons learned from early mistakes.

Strengths of the Counterinsurgents

Northern Command has since been heavily involved with additional civic and developmental operations. Following the success of its initial large-scale operation, the Indian government approved funding for additional projects.

In the early years following the end of the World War II, the Indian Army composed a doctrine for COIN operations, using as a foundation the British experience in Malaya. Since then, the doctrine has been refined by putting theories to practice and adapting them to allevi-

[12] Sood and Sawhney (2003).

ate problems common across the spectrum of operations. Further lessons were learned through the army's experiences against the Liberation Tigers of Tamil Eelam (LTTE) in Sri Lanka and the North-East insurgencies.

The Indians quickly recognized that the very nature of COIN operations is one of constant unpredictability. Lieutenant General Ray writes:

> The only certainty is uncertainty. Low-intensity conflict is all about high-speed change, chaos, and disorder. It demolishes in one stroke all traditional military concepts applicable to general war. The contradictions are simply too many.[13]

To successfully and consistently combat insurgent forces, COIN forces must create and maintain a secure environment that can be regulated with relative ease. This requires the deployment of security forces across the region. According to Indian doctrine, there are five important steps must be taken to successfully conduct COIN operations:

- separation of civilians from insurgents
- the use of a linear grid system
- physical domination of an area of responsibility (AOR)
- restraint in use of airborne and land-based firepower
- civic action (winning the hearts and minds).

The first requirement to successfully carry out COIN operations is to isolate the insurgents from the local population by temporarily shifting the civilians to villages already secured by Indian forces. This better allows security forces to effectively screen the locals for insurgents.

As the threatened regions of J&K are vast and require large numbers of security personnel, the Indian Army has regulated its COIN operations through the use of a grid system. An army battalion's AOR is essentially demarcated by the level of insurgent activity and the ease with which operations can be carried out (because of such factors as terrain and size of local population).

The large grid can then be further divided into smaller sections that can then be monitored easily by a company- or platoon-sized unit of soldiers. If necessary, fire support can also be called in to deal with larger insurgent groups. As COIN operations are often conducted on the platoon level (or sometimes even squad level), the implementation of such a system helps to reduce gaps and allows a unit not only to observe its AOR more consistently but also to dominate it.

With a strong physical presence in an area and vigorous patrols conducted during both day and night, security personnel are able to observe and regulate a given village and the access it has to its surroundings. Soldiers cultivate human intelligence (HUMINT) through a cache of local assets and agents. Direct familiarity with the villagers can help soldiers detect the presence of any unusual or suspicious persons.

---

[13] Ray (1997).

Additionally, the Indian Army made a conscious decision to severely curb the use of airborne and artillery-based firepower while conducting COIN operations to minimize collateral damage and corresponding casualties. The excessive escalation of firepower during firefights has severely disrupted the local population and caused a much higher unintended casualty rate. Two unsuccessful COIN campaigns—the U.S. military in Vietnam and the Soviet Army in Afghanistan—yield examples of the negative consequences that result from the continued overuse of heavy firepower in civilian areas.

Forces carrying out COIN missions must have a formidable resource of men and equipment, as this sort of restraint can be costly. In close-quarters combat, casualties to the COIN forces will be unavoidably higher than if using distanced artillery strikes or air support.

In addition to traditionally military operations, it is of vital importance to actively engage and promote interaction between the security personnel and the civilian population. Lieutenant General S. C. Sardeshpande reaffirms that "counter insurgency operations must, of necessity, be an intimate mix of military operations, civic actions, psychological operations, and political/social action."[14]

Good media relations can also effectively "showcase" COIN operations, providing increased comfort to the local population and even intimidating insurgent groups. Active and continuous interaction with both local and international media groups can serve as a force multiplier. Keeping the population informed can only help to alleviate any alienation within it.

Recently, 15 Corps of the Indian Army launched an expensive and ambitious project, Operation Sadbhavana. The brainchild of Lieutenant General Arjun Ray, it was a large-scale venture aimed at improving life for the civilians of Ladakh in J&K. With an initial cost of close to $1 million, the plan called for constructing schools, hospitals, and community development centers and providing water and electricity. The project also included tours for locals to different parts of the country and the improvement of roads and bridges across the state. The project was widely acclaimed and declared a success.[15]

It is important to note that these civic actions were conducted on a significant scale and in a transparent, genuine manner. The government determined that worthwhile facilities for the population had to be properly planned and initiated, with proper follow-through to start winning hearts and minds. A clear and visible difference in the lives of the locals was also needed if they were to begin trusting security personnel. Otherwise, the feelings of alienation would foster further anti-Indian sentiment.

Anit Mukherjee, who served in the Rashtriya Rifles in Kashmir and Nagaland, writes:

> After the first year of conducting operations with questionable results, my unit made a significant shift toward people-friendly operations. That meant taking off shoes before searching mosques, deciding not to search old men, women and children and even letting insur-

---

[14] Sardeshpande (1993).

[15] The guidelines used for Operation Sadbhavana are currently being adapted for implementation in other areas affected by insurgency, including other regions of J&K and the Indian Northeast. For the success of the operation and for his service in Ladakh, Lieutenant General Ray was awarded the Param Vishist Seva Medal, the highest award in the Indian military, for distinguished service.

> gents escape rather than risking a firefight in a built-up area. Over time, our hard work paid off. Tips became more frequent and reliable. As we gained the trust of the locals, we succeeded in preventing recruitment while eliminating insurgents.[16]

In recent years, the amount of violence in J&K has decreased, largely because of a substantial shift in Indian COIN strategy. However, insurgent groups continue to enjoy sanctuary in Pakistan. With the help of the ISI, the insurgents are able to rearm, train new recruits, and then redeploy into J&K. The border is still too porous; Indian security forces are simply unable to guard the entire stretch of the LoC from infiltration. Military success, regardless of how extensive, can never defeat an insurgency. Indian soldiers, as part of their indoctrination into COIN operations in the state, are told that the insurgency will end only when an effective political solution is developed and adopted; military action alone is not enough. Until that time, the violence across J&K can only be *managed* by security forces.

---

[16] Mukherjee (2006). Mukherjee served in the Indian Army for nine years and attained the rank of major. He currently is a doctoral candidate at the School of Advanced International Studies at Johns Hopkins University.

CHAPTER SEVEN

Colombia (1963–Present)

Angel Rabasa

Origins and Characteristics of the Insurgency

Insurgency has been endemic to Colombia. For most of the country's history, political transitions came about as the result of successful insurrections by the party out of power. The 20th century saw two major civil wars or periods of extreme political violence: the War of the Thousand Days (1898–1900) and the period known as *La Violencia* (literally, the violence), from 1948–1953, which is estimated to have resulted in 300,000 deaths. *La Violencia*, and the subsequent military dictatorship of General Gustavo Rojas Pinilla, were brought to a close by the National Front in 1957, in which the country's two dominant parties, the Liberal and Conservative parties, agreed to end their confrontation and alternate in power for the following 16 years. (See Figure 7.1 for a map of Colombia.)

This context is necessary to understand the development of the modern Colombian insurgent movements—primarily the Revolutionary Armed Forces of Colombia (FARC) and the National Liberation Army (ELN).[1] The Cuban revolution and the beginning of Soviet support for national liberation movements in Latin America encouraged the advocates of revolutionary change in Colombia to challenge the National Front governments through armed struggle. In 1961, the Colombian Communist Party declared, "the revolutionary path in Colombia could require a combination of all forms of struggle." The armed struggle was launched on a foundation of experienced fighters. The FARC leader, Pedro Antonio Marin, better known by his nom de guerre Manuel Marulanda Vélez or *Tirofijo* ("Sureshot"), for instance, began his guerrilla career in 1949 when he joined a Liberal guerrilla band in the department of Tolima, one of the epicenters of the violence.

In 1964, Marulanda helped to establish a communist-oriented "independent republic" in Marquetalia, a remote area in southern Tolima. This was one of several "republics" established by communist-oriented guerrillas in southern Colombia. Marulanda escaped when the Colombian Army attacked and destroyed the Marquetalia guerrilla group in June 1964. In

[1] Numerous minor guerrilla groups developed in the early stages of the Colombian insurgency, for instance, the Chinese-backed Popular Liberation Army (ELP), the ERP, the Guevarista Revolutionary Army (ERG), the M-19 Guerrilla Movement, and others.

Figure 7.1
Colombia

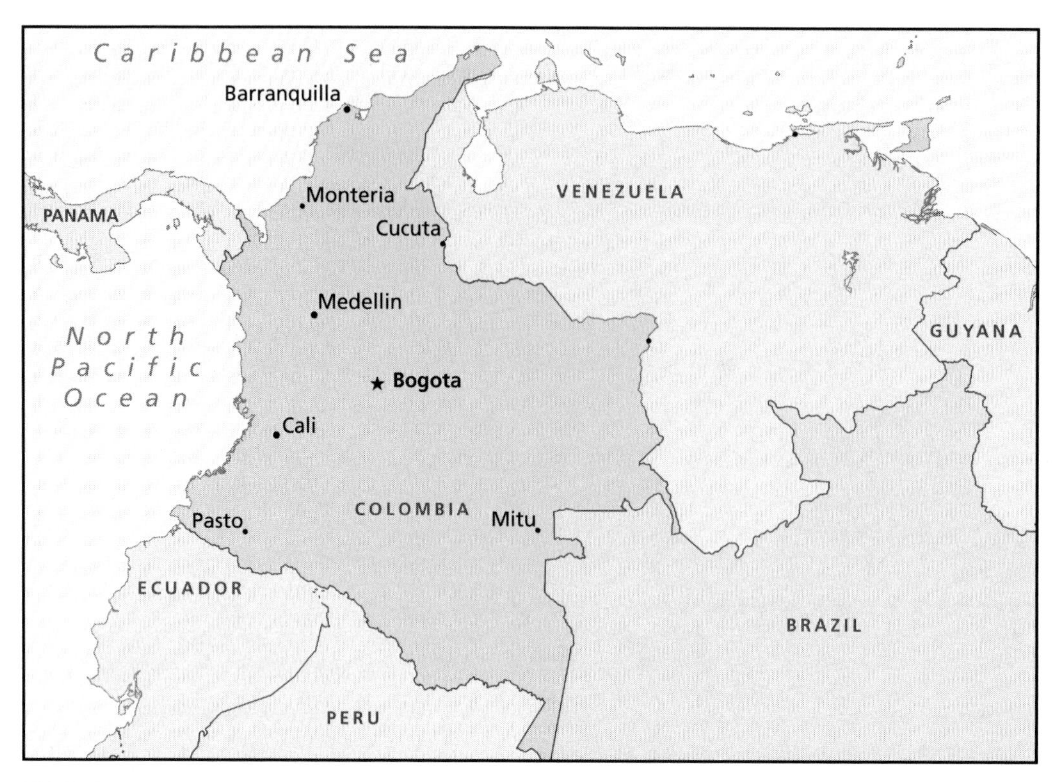

RAND *OP185-7.1*

1966, the communist guerrillas, denominated "self-defense forces," were reorganized as FARC, with Marulanda as military commander.[2]

The FARC expanded slowly between the mid-1960s and the 1980s. In its early stages, FARC guerrillas engaged in ambushes of military units and raids on farms. The main objectives were capturing military equipment, securing food and supplies, capturing hostages, and settling scores with informers. In this early stage, the FARC was more concerned with survival than with expansion.[3] In 1974, the organization established a general staff and a secretariat to provide political direction. Nevertheless, during this period, the FARC operated only in remote, marginal areas of the country and did not pose a serious threat to stability.

This situation changed in the 1980s, when the FARC began to implement the expansion strategy set at its 1982 Seventh Conference, funded by resources captured from the illicit drug trade. Until that time, the FARC, like other revolutionary organizations, was reluctant to become involved in the drug trade, which it considered counterrevolutionary.[4] With the

---

[2] The FARC officially dates its foundation to 1964, so the organization, the oldest guerrilla movement in Latin America and perhaps the world, has been active for 42 years as of this writing.

[3] Maullin (1973).

[4] See Rabasa and Chalk (2001, p. 26).

revenue from the drug trade (initially by taxing it), the FARC began to implement the strategic and operational concepts outlined by the Seventh Conference: first, to consolidate control of coca-growing regions; second, to extend the theater of operations to the entire country, so as to force the government to disperse its forces and reduce its ability to regain the military initiative; third, to isolate the capital, Bogotá, and other major cities; and fourth, to transition to large-scale military operations, culminating in a general uprising.[5] The effectiveness of the FARC's strategy and operations can be assessed in terms of its success—or lack thereof—in attaining these objectives.

The second group considered here is the ELN. Unlike the FARC's rural-origin leadership, the ELN leadership was composed of university students who followed the Havana line. The ELN also attracted radicalized Catholic adherents of liberation theology. The ELN suffered severe attrition in the late 1960s as the result of government pressure and internal purges, but it reemerged in the 1980s under the leadership of a guerrilla priest named Manuel Pérez (who died in Havana in 1998). The ELN has been more reluctant than the FARC to exploit the illegal drug trade. Kidnappings and extortion (including protection payments extracted from the oil industry) account for most of the ELN's income, but some more "pragmatic" sectors of the group have become involved in the drug trade.

Strengths of the Guerrillas

The FARC and the ELN are some of the longest-lasting insurgencies in the world. Over a period of 40 years, they have survived efforts by the Colombian military to eradicate them and have expanded their presence throughout the country.[6] Some of the factors that account for the guerrillas' survival are the following:

1. *Strategic flexibility*. The FARC and ELN have successfully adapted over the years to changed circumstances, while maintaining (especially the FARC) a consistent strategy of expanding their presence throughout the country. The FARC has expanded from 350 fighters at its formal establishment in 1966[7] to 32 fronts with 3,600 fighters in 1986, 60 fronts with 7,000 fighters in 1995, and over 70 fronts with 15,000–20,000 fighters in 2000.[8] When the government has gone on the strategic offensive—for instance, during the mid-1960s Plan Lazo offensives and after the launching of the Uribe government's military strategy, known as Plan Patriota—the FARC has gone on the defensive and lowered its military profile, to resume attacks at a more favorable time later. The ELN

[5] Fuerzas Armadas Revolucionarias de Colombia (1994).

[6] The most serious of these efforts was Plan Lazo, developed by the Colombian armed forces in consultations with U.S. counterinsurgency planners in the early to mid-1960s. Plan Lazo failed to eradicate the guerrillas but succeeded in eliminating a number of guerrilla groups and in extending government control to many contested areas. See Maullin (1973, pp. 69–80).

[7] The FARC dates its foundation to 1964 as the Southern Bloc self-defense forces; it was renamed FARC in 1966.

[8] Rangel (1998, p. 12).

also experienced significant growth in the 1980s and 1990s but not as much as the FARC. It attempted to expand to other regions of the country, but by and large the group's fighting force has also remained concentrated in its traditional area of influence in northeastern Colombia.

2. *Unity of command and organization.* Almost since the FARC's inception, Marulanda has dominated the organization, first as chief of staff and subsequently as head of the secretariat. There have been no serious challenges to the leadership throughout the history of the organization. The FARC is hierarchically organized with a seven-member secretariat, which provides political direction to the movement and supervises a 25-member general staff and the chain of command for the People's Army. There is some question, however, as to the extent to which the secretariat controls the 70-odd largely self-financing fronts dispersed throughout the country, particularly after the loss of the "demilitarized zone" that the FARC controlled between 1999 and 2002. In contrast to the FARC, the ELN has experienced significant dissension, particularly in its early years, when the group nearly disintegrated because of infighting and Colombian military pressure. The most recent purge in 2002 resulted in executions and desertions.[9]
3. *Expanded sources of income.* Estimates of the FARC's income range from $200 million to $400 million[10] to over $1 billion[11] annually, half of it from the drug trade. According to Latin America security analyst Samuel Logan, the old system of limited arms purchases with cash has been replaced by arms-for-cocaine swaps, using the same routes used to smuggle cocaine out to smuggle the guns in. The arms suppliers include Mexican and Brazilian organized crime networks.[12] The changed dynamic produced by the drug income is illustrated by the minutes of the FARC secretariat's annual meeting, published by the Colombian weekly *Semana* in August 2000: "The acquisition of arms currently has permitted us a qualitative jump in our process of becoming the Ejército del Pueblo [People's Army or FARC-EP]."[13] As noted above, the ELN has entered the drug trade partially and more reluctantly than the FARC. (Reportedly, coca-related income currently amounts to one-third of the ELN's revenues.)[14] The group relies for much of its income on kidnappings and extortion—particularly payments extracted from the oil and mining industries in areas where the ELN is active.[15]
4. *Exploitation of poorly controlled borders.* Narcotraffickers and guerrillas have developed an extensive infrastructure that enables them to move drugs, arms, supplies, and personnel across Colombia's permeable borders with Venezuela, Panama, and Ecuador, where

[9] Espejo and Garzon (2005).

[10] Council on Foreign Relations (2005).

[11] Colombian armed forces briefing, 2000.

[12] Logan (2006).

[13] "Los Planes de las Farc" (2000).

[14] Espejo and Garzon (2005).

[15] These areas include, particularly, the oil production facilities in Arauca department and the Caño Limón pipeline, the Barrancabermeja petrochemical complex, and gold mining enterprises in Antioquia.

government presence is limited to the immediate areas around population centers. This infrastructure has been key in sustaining the FARC and ELN. These areas constitute a strategic rear guard that enables FARC and ELN forces to resupply and slip across the border when under pressure by government forces. The Venezuelan border is particularly problematic because Venezuelan President Hugo Chávez has ignored FARC and ELN cross-border activity and tolerates their presence and activities in Venezuela.[16]

Weaknesses of the Guerrillas

The weaknesses and shortcomings of the FARC and ELN, if anything, are greater than their strengths and make it highly unlikely that either group could overthrow the Colombian government in the foreseeable future. These weaknesses include

1. *The downside of involvement in the drug trade.* Although the drug trade produces high profits, it is also an important source of weakness. For the FARC, the effects have been (a) loss of ideological cohesion, particularly at the lower levels, as some of the commanders have become more interested in profits from the drug trade than in advancing the organization's political-military agenda; (b) loss of domestic and international support, as many potential supporters were repelled by the FARC's involvement in obviously criminal activity; and (c) facilitation of the building of coalition in support of the Colombian government.
2. *Failure to gain popular support.* Although it has some level of support (willing or coerced) in its base areas, the FARC and ELN remain deeply unpopular with the Colombian public. The FARC and ELN have never registered higher than 2 or 3 percent in popularity polls. The guerrillas' resort to kidnappings and involvement in other criminal activity, attacks on economic targets, and indiscriminate attacks on civilians—for instance, the May 2002 attack on a church in Bojayá which killed 119 persons—all contributed to their loss of popular support.
3. *Failure to sustain the move from small-unit attacks to large-unit operations (battalion size or larger).* In the late 1990s, the FARC attempted to make a qualitative jump to a higher stage of military operations by engaging and defeating battalion-sized units of the Colombian Army (Las Delicias in 1996 and El Billar in 1998).[17] However, the sequence of successful large-scale attacks on isolated army units was broken at the battle of Mitu (November 1988) where, for the first time, the Colombian military used on-call air power to defeat a FARC attempt to overwhelm the local garrison. Since then, the Colombians have used air-land synergies to prevent guerrilla concentrations. The ELN has never attained the ability to mount large-scale operations. Over the last several years, in fact, the organization has come under pressure by the FARC and paramili-

[16] Robinson (2003).

[17] Rabasa and Chalk (2001, pp. 42–43).

tary forces, which have impinged on its territory, and it is now largely concerned with survival.[18]

4. *Failure to mount sustained attacks on strategic centers.* Although in the 1980s and 1990s the FARC carried the armed struggle beyond the traditional guerrilla areas in periphery (and even mounted attacks near Bogotá in 2000), it failed to challenge the government's control of the country's populated heartland. There is a mixed record with regard to economic warfare. The FARC and ELN have attacked the transportation, communications, and energy infrastructure—interdicting highways, destroying electricity towers, and attacking oil pipelines. However, this activity has not disabled vital economic assets. In 2004, the Colombian economy was one of the best performing in Latin America, with an annual growth rate in gross domestic product of 3.5 percent.
5. *Failure to deploy sophisticated weapons.* The FARC and ELN have been very ingenious in using locally available materials to develop improvised weapons, such as homemade propane cylinders filled with explosives launched from improvised mortars fashioned from 55-gallon drums or larger canisters. Nevertheless, despite large revenues and links to international illegal arms networks, the FARC has not been known to deploy sophisticated infantry weapons, especially surface-to-air missiles, which could neutralize the Colombian government's air superiority, a critical component in the government's military edge over the FARC.
6. *Loss of the demilitarized zone.* As a condition for its participation in peace negotiations beginning in January 1999, the government withdrew its forces from the *zona de despeje*, an area of some 42,000 square kilometers in south-central Colombia. The FARC established its headquarters in this zone and used it as a sanctuary to launch operations, rest and refit its forces, and hold prisoners and hostages. The zone was the hub of a series of logistical and mobility corridors that the FARC used to move fighters, arms, and drugs throughout the country. After three years of fruitless talks, however, the Pastrana government forces reoccupied the *zona de despeje* in January 2002. The proximate cause was the spike in FARC attacks in the preceding months, highlighted by the hijacking of an airliner and the kidnapping of a Colombian senator. The underlying reason was the failure of the peace process to produce any results—largely because of FARC intransigence—and the FARC overplaying its hand by taking ever more provocative actions.
7. *Political weakness.* The loss of its sanctuaries in the *zona de despeje* represented a deeper political failure on the part of the FARC. The FARC's strategy of simultaneously fighting and talking created tactical advantages for the group but shifted the political dynamics in Colombia in favor of advocates of a hard line toward the guerrillas—as shown by President Alvaro Uribe's election with an unexpectedly large vote in May 2002 and his reelection in 2006 with an even larger majority.

[18] The loss of Barrancabermeja—a major source of ELN income—to the paramilitaries in 2000–2001 was a particularly hard blow to the ELN (Espejo and Garzon, 2005).

Strengths of the Government

The strengths and weaknesses of the government side are, in some respects, the obverse of the guerrillas' strengths and weaknesses. The main strengths of the Colombian government in its contest with the various antistate forces are the following:

1. *The democratic character of the Colombian state.* Although imperfect, Colombia has one of the longest records of elected civilian governance in the continent (there was only one military regime in the 20th century, that of General Rojas Pinilla, 1953–1957). Since the end in 1974 of the alternation in the presidency of the Liberal and Conservative parties under the National Front arrangement, elections have been free and competitive. The preservation of constitutional government even under conditions of insurgency and high levels of violence has been a key factor contributing to maintaining the government's legitimacy and domestic and international support.
2. *Popular and competent leadership.* As the case of El Salvador also shows (see the discussion of the role of President José Napoleón Duarte), competent and credible leadership is critical in rallying a government under attack to defeat an insurgency. President Uribe is regarded as a strong and competent leader and his hard line toward the guerrillas generated unprecedented approval rates of up to 70 percent.[19] Uribe's popularity and high regard for his leadership led to the amendment of the Colombian constitution that permitted him to run for reelection. He was reelected in 2006 with 62.2 percent of the vote, a margin of 40 percent over his nearest rival.
3. *International support.* Improvements in Colombia's human rights performance and the damage to the FARC's image as a result of its involvement in the drug trade have helped to maintain strong international support for the Colombian government. Colombia is the third largest recipient of U.S. assistance (not counting Iraq and Afghanistan). The European Union (EU) also supports the Uribe government's fight against terrorism and drug trafficking[20] and provided some €300 million in assistance to the Colombian government, local governments, and nongovernmental organizations (NGOs) for the years 2001–2006.
4. *Successful adjustment to insurgent strategy and tactics.* Since the mid-1990s, when the FARC was able to engage and defeat battalion-size Colombian army formations, new operational and tactical approaches on the part of the Colombian military (including coordinated air-ground operations) have produced improved military performance. Under Uribe, the Colombian government inaugurated a seize-and-hold strategy, which substantially increased the territory under government control and expanded overall government forces strength by over 80,000, to 207,000 active military personnel and 129,000 police and other security forces (facing an estimated 12,000 to 15,000 FARC

[19] "Colombia President Rides Popularity Wave: Uribe Widely Seen as Tough-Minded" (2005).

[20] The FARC, ELN, and AUC are all included in the EU's list of terrorist organizations

and some 3,000 ELN fighters).[21] The Uribe government also inaugurated a program of training and arming local self-defense forces under military supervision.

Weaknesses of the Government

Despite notable advances over the last several years, there are weaknesses in the Colombian institutional and military structures that could be exploited by the FARC if the Colombian government's current efforts faltered. These include

1. *Uribe's indispensability.* The downside of Uribe's hands-on style of leadership is that he has become indispensable. Although his leadership has been instrumental in energizing the Colombian government's response to the FARC threat and rolling back many of the gains made by the guerrilla organizations in the preceding decades, there is no obvious successor with Uribe's charisma and popularity. Much of the momentum currently on the government's side could dissipate if he were to be removed violently (there have been several assassination attempts) or when his second term runs out in 2010.
2. *Preservation of FARC capabilities and infrastructure.* Despite the progress made by the government forces in clearing strategic areas, such as the region around the capital, the FARC retains a presence in almost all of Colombia's departments, including coca-growing areas in eastern and southern Colombia. The FARC's military capabilities also remain substantially intact. Some Colombian analysts suspect that the FARC might have implemented a tactical retreat to build up its strength and regain the initiative at a later time.
3. *Persistence of criminal networks.* Criminal networks pervade the Colombian landscape. These networks traffic drugs, weapons, and other illegal goods and engage in money-laundering, kidnapping, and extortion, sometimes in league with the FARC, ELN, and AUC. The activities of these groups greatly complicate the government's task of fighting the guerrillas and consolidating state control.
4. *Inability to control borders.* The Colombian government has had great difficulty in creating a secure environment on the borders with Panama, Venezuela, and Ecuador. The FARC and, to a lesser extent, the ELN maintain a significant presence and influence in the border regions as well as sanctuaries on the Venezuelan and Panamanian sides of the border. The government's dilemma is that it does not have enough forces to clear and hold territory in the country's heartland and simultaneously secure the borders.
5. *Vulnerability to external interference.* There is strong evidence that people in Venezuelan President Hugo Chavez's inner circle have provided support to the FARC—ammunition, safe houses, documentation, and even weapons. Moreover, Venezuela's acquisition of sophisticated military equipment from Russia and other sources will give Venezuela an overwhelming military superiority over Colombia. Even though Colombian analysts do not believe that Chavez is preparing for war with Colombia, at a min-

[21] International Institute for Strategic Studies (2006).

imum, the new Venezuelan armaments cannot but distract the Colombian military from their primary counterinsurgency mission and relieve pressure on the FARC.

Conclusions

Historically, Colombian governments, for the most part, considered the existence of the guerrillas as a fact of life that could be ignored as long as the guerrillas did not constitute a strategic threat to the stability of the state. This was a viable approach until the convergence of the insurgency and the drug traffic gave the FARC the means to grow and expand into economically and politically strategic regions of Colombia. The Colombian government's response was to attempt to negotiate with the guerrillas. A truce with the FARC and parts of the ELN was in effect from 1984 to 1987 and peace negotiations took place under the administrations of César Gaviria (1990–1994), Ernesto Samper (1994–1998), and Misael Pastrana (1998–2002). All of these efforts to reach a negotiated settlement failed because the guerrillas were not seriously committed to reaching a peace agreement but, rather, saw the talks as a way to gain tactical advantages in a long-term campaign to seize power.

The United States, for its own part, recognized the nexus between the guerrillas and drug trafficking but has failed to identify the political control that the guerrillas exercise over an ever-larger part of Colombia's territory as the center of gravity of the guerrilla/drug trafficking complex. As a result, U.S. efforts were focused on strengthening Colombian antinarcotics capabilities while insisting that U.S. military assistance not be used for counterinsurgency. This was unrealistic, given the Colombian government's inability to eradicate the drug trade where it did not have physical control and the magnitude of the political and military threat that it faced from the growing strength of the guerrillas.

In short, throughout the 1990s, neither the Colombian government nor its U.S. ally understood the nature of the security problem in Colombia. This strategic myopia began to be corrected with the breakdown of the peace process in the last year of the Pastrana administration and the election of President Uribe in 2002 and, for the United States, with the events of September 11 and the onset of the U.S.-led Global War on Terror.

The Uribe government's counterinsurgency approach, embodied in the Defense and Democratic and Security Policy and its military component, Plan Patriota, centered on consolidating control of territory and protecting the population—classical counterinsurgency doctrine. In addition to increasing defense spending by 30 percent and troop strength by over 80,000 personnel between 2001 and 2004, the Uribe government has begun to build civil defense units in rural areas under military supervision. For its own, the United States changed its policy that previously limited the use of U.S.-provided equipment to antidrug operations, freeing it up to be used for counterinsurgency operations. The counterinsurgency effort in Colombia is now therefore based on a more adequate strategy and is being carried out more forcefully than at any time since the 1960s.

Although it is too early to make a conclusive assessment of the current Colombian approach, there are indications that the government is meeting its objectives. The government has established a presence in every one of Colombia's nearly 1,100 municipios (in 2000, there was no police presence in one-fourth of the country's municipios)[22] and garnered significant successes against the FARC. Attacks by "irregular groups" (as the guerrillas and paramilitaries are denominated) on government forces in 2005 diminished by 9 percent from 2004 and by 29 percent from 2003.[23] Of course, no one expects that the government will be able to eradicate the guerrillas any time soon. Rather, the expectation is that the new Colombian strategy and tactics will gradually reduce the scope of the guerrillas' activities and create the conditions for realistic negotiations.

---

22 Colombian armed forces briefing, 2000. Municipios are the second level of local administration, after the departments (currently 32). They are the equivalent of U.S. counties or parishes.

23 Fundación Seguridad y Democracia (2006). On the other hand, Colombian security analyst Alfredo Rangel notes that the FARC's military capabilities remain intact in many areas and that the process of demobilization of the paramilitaries promoted by the government could paradoxically increase the FARC's freedom of action (Suárez, 2005).

CHAPTER EIGHT

Conclusions: Lessons Learned for Future Counterinsurgencies

Lesley Anne Warner

In an age when insurgencies have worldwide reach, counterinsurgents can ill afford *not* to examine the complexities of past cases and the continuities among them, especially since the complexities of the insurgency that the counterinsurgents are facing may not be elucidated until much later. There is more to gain than to lose by examining past insurgencies; as insurgents learn from other insurgencies, counterinsurgents should continue to learn from the successes and mistakes of other counterinsurgencies to avoid the repetition of mistakes. This may especially be true as the United States continues to confront a variety of mutating threats in several parts of the world. However, in thinking of future insurgency threats, the most valuable lesson is that COIN successes and failures are loose analogies. The insurgencies of the past differ somewhat from the insurgencies the United States faces today in its Global War on Terror, although some characteristics, such as the importance of eliminating sanctuary and the importance of providing security for the population, have not changed.

In the following paragraphs, the lessons learned from the COIN operations in this paper and other cases can be classified by what the recently released U.S. Army and U.S. Marine Corps Counterinsurgency Field Manual (FM 3-24/ MCWP 3-33.5) describes as Defensive and Offensive Stability Operations: Civil Security, Civil Control, Essential Services, Governance, and Economic Infrastructure and Development.[1] Because of counterinsurgency's complex and multifaceted nature, the author finds it insufficient to compartmentalize what in some cases is a lesson that spans two or more of these categories. Unless specified, these lessons are applicable to host nation counterinsurgents as well as to foreign counterinsurgents, as the situation permits.

According to the Counterinsurgency Field Manual (FM 3-24/ MCWP 3-33.5), "an effective counterinsurgent force is a learning organization."[2] Successful COIN requires unfettered adaptability and the ability to be an objective critic in the face of failure.[3] The emphasis here is on the ability to learn from failure, as success in COIN must not be seen as unequivocal. As Cohen et al. argue, if a program is successful in one province or city, it will most likely be

[1] Department of the Army and U.S. Marine Corps (2006, Section 1-19).

[2] Department of the Army and U.S. Marine Corps (2006, Section 1-26).

[3] Metz and Millen (2004, p. 26).

unsuccessful in another.[4] Similarly, if a solution is successful in one counterinsurgency, it may be unsuccessful in another, possibly because of intervening variables. The reason for this paradox is that conditions differ in several places and, thus, solutions must be custom-made for each situation. Although the insurgency is still ongoing, Indian counterinsurgents in Jammu and Kashmir have learned which tactics, techniques, and procedures (TTPs) bring success and which bring failure and have adjusted their strategy accordingly.

Adaptability to insurgent tactics and the ability to learn from the effectiveness or lack thereof of other COIN campaigns proved an advantage in the Philippines, Vietnam, Jammu and Kashmir, and Colombia. In these campaigns, counterinsurgents exhibited varying degrees of willingness to learn from the U.S. Indian Wars, the British experience in Malaya, the Indian campaigns against the LTTE in Sri Lanka, and Colombia's own long experience against insurgents.

Granting counterinsurgents local autonomy would hasten the process of innovation and adaptability, which should be encouraged and accepted at all levels of bureaucracy.[5] Rather than having a bureaucrat thousands of miles away making general decisions for all theaters of the insurgency, local authorities who have their fingers on the pulse of the population can assess the situation on the ground and then determine how best to combat local problems.[6] In the case of J&K, the counterinsurgents have divided their AOR into grids, which has allowed military units to become intimate with the particulars of their environment, as well as reduce gaps in the provision of security.

Military and civilian agencies must work together to create innovative and adaptive COIN programs and tactics. Within the unity of effort, the military must secure inflamed areas so that political and civic actions can follow, building on the success of the security forces. The level of interaction between the civilians and the military should be such that the left hand knows what the right hand is doing and how cooperation can facilitate the success of operations. If there is not enough communication between civilians and the military, bureaucratic competition and deadlock will set in, thereby undermining the COIN objectives of both sides. Such bureaucratic cooperation should take place not only in the conflict zone but in the seats of government of both the host nation and the foreign counterinsurgents so as to prevent cleavages for the insurgents to exploit.[7] In this effort, it is essential that bureaucratic self-interest not have a detrimental effect on the execution of COIN. In this paper, the case of Vietnam is an excellent example of how a bureaucratic impasse impeded successful COIN in the early years of the insurgency.

The population may be more likely to accept the presence of foreigners if it sees that they are contributing to progress rather than chaos.[8] In the Philippines, Algeria, Vietnam, and J&K, the counterinsurgents engaged in humanitarian efforts relating to health, education,

[4] Cohen et al. (2006, pp. 49–53).

[5] Metz and Millen (2004, p. 25); Nagl (2002).

[6] U.S. Department of the Army and U.S. Marine Corps (2006, Section 1-26).

[7] U.S. Department of the Army and U.S. Marine Corps (2006, Section 1-22).

[8] U.S. Department of the Army and U.S. Marine Corps (2006, Section 1-23).

infrastructure, and the protection of basic human rights to improve the lives of the population whose acquiescence they sought. By providing the funding and the guidance for such activities as building schools and digging ditches for waste management, counterinsurgents can make employment available so that laborers can do honest and productive work toward stabilizing the political and economic situation while providing for their families and stimulating the local economy. Giving the population a way to support itself and showing that counterinsurgents are instruments of change for the better in the process of stabilization gives potential insurgents a stake in the process of political and economic development.[9]

Civic and humanitarian actions, along with credible pledges of protection in return for cooperation, can help the counterinsurgents gather HUMINT, although in the cases of Algeria and Vietnam, these efforts came too late to be effective or were undertaken on such a small scale that the positive effect was minimal. In the case of the Philippines, humanitarian efforts were more along the lines of carrot and stick, in which the population was presented a rational choice to either support the counterinsurgents and gain rewards or support the insurgents and face punishment.[10]

Information operations (IO) should be employed to publicize actions that exemplify host nation legitimacy and competence, manage the gap between the population's expectations and the counterinsurgents' ability to respond to their needs, and discredit insurgent propaganda.[11] If information is made available to the population explaining how counterinsurgent actions have increased personal security and improved conditions in the "conflict ecosystem," the population may be more inclined to support the counterinsurgents.[12] To ensure maximum availability of disseminated information, counterinsurgents should use culturally appropriate symbols in multiple forms of communication, taking into account which means are most accessible to the population in question. Much like other aspects of COIN, the best IO will be localized, since circumstances differ from city to city and province to province.

The Counterinsurgency Field Manual (FM 3-24/ MCWP 3-33.5) posits that, for the COIN effort to attain lasting success, the host nation needs to achieve legitimacy.[13] Politically, the presence of strong, competent, democratically elected leaders in El Salvador and Colombia allowed the regimes to maintain legitimacy and gave the population a viable outlet for possible grievances. To establish or maintain political legitimacy, counterinsurgents must be willing to listen to various groups and integrate them in the political process so that they have a stake in the future of their country. Counterinsurgents must sometimes even be willing to listen to the grievances of some of the less-agreeable elements of society, as a permanent end to the conflict

---

[9] Sepp (2005, p. 9).

[10] In instances where the insurgency is of a military nature, as in Jammu and Kashmir, having counterinsurgents carry out actions to gain the trust of the population or gain intelligence from it, can help more than it can hurt, as the key to ending an insurgency is often persuading those in the population that they have the opportunity to have a better life by siding with the counterinsurgents.

[11] U.S. Department of the Army and U.S. Marine Corps (2006, Section 5-2).

[12] "Conflict ecosystem" is a term used by David Kilcullen to describe all aspects of the environment in which the insurgency takes place (Kilcullen, 2006–07, pp. 111–130).

[13] U.S. Department of the Army and U.S. Marine Corps (2006, Section 1-22).

cannot be established unless the various warring factions can see their places in the future of the country. In El Salvador, a negotiated settlement gave insurgents incentives to lay down their arms and address their grievances through the political process. In pursuit of legitimacy and endurance, it is important that such agreements be timed so as to minimize the issue of foreign occupation.

Despite providing foreign aid for COIN in El Salvador, the United States placed a ceiling of only 55 military personnel in the country throughout the insurgency, unlike the gradual transition from advisers to combat troops throughout U.S. involvement in Vietnam. Conversely, the presence of foreign combat troops, as in the Philippines, Algeria, and Vietnam, fanned the flames of nationalism on the insurgent front. In two of these cases, the insurgency emerged as a way to resist foreign occupation, which is similar to the insurgencies the United States is facing today. Since every army of liberation has a half-life according to General David Petraeus, foreign counterinsurgents or even host nation counterinsurgents who are not from that particular area of operations should not expect much latitude, making it critical for an expedient return of sovereignty and authority to the host nation in all aspects of COIN.[14]

Deciding to help a host nation that is facing an insurgency brings up the issue of leverage. As Dan Byman argues, foreign counterinsurgents need to carefully determine how best to effectively bridge the gap between their interests and those of the host nation. As it is, the more critical a situation is to U.S. national security interests and the more the United States pledges its support for the host nation, the less leverage it has over how its aid money is spent and how the war is fought.[15] To mitigate this common difficulty, Byman suggests that foreign counterinsurgents diversify their sources of intelligence, possibly using NGO status reports and open-source media to augment the volume and quality of information they are being provided, as well as consider gathering information on the ally's intelligence collection and dissemination activities.[16]

In contemporary counterinsurgencies, we are learning that motivating the population to choose sides is more critical than in the past because of the speed at which information is made available to target populations from either the insurgent or counterinsurgent side. Using indigenous human capital, counterinsurgents should aim to build trust with the population and understand the "conflict ecosystem" through established social networks,[17] as they should already be endowed with the knowledge of how best to earn the trust of the population.

The use of indigenous forces, especially forces from the particular area in question, increases the legitimacy of the counterinsurgents and can also help to divide and weaken the insurgency by psychologically unhinging the insurgents. In the Philippines, Algeria, Vietnam, El Salvador, and Colombia, the counterinsurgents recognized the advantage of arming and training trustworthy civilians to defend their villages, which allowed the military more time to pursue the insurgents.

---

[14] Petraeus (2006, p. 4).

[15] Byman (2006, p. 81).

[16] Byman (2006, p. 112).

[17] David Kilcullen, quoted in Packer (2006, p. 64).

The use of indigenous intermediaries would help bridge the gap between foreign counterinsurgents and the population, especially if there are moderates who may be more willing to listen to what each side has to say. Moderates may be targets of the insurgency because of their perceived proclivity to compromise with the host nation. As in the case of Algeria, counterinsurgents should take advantage of the existence of such a population and seek to expand its ranks while offering strong and credible protection, as these activists are more likely to become targets of the insurgency.

Enhanced language abilities and cultural training would allow trust to be built at a faster pace—on both the political and military fronts of the insurgency. Once those in the population see that the counterinsurgents have made a genuine effort to learn their language and understand their culture, they may become more amenable to pacification.

Understanding local dynamics will help put both civic and military actions in perspective, as the population can hopefully impart wisdom on local conditions existing prior to and during the insurgency, allowing the counterinsurgents to better comprehend the general trend of the insurgency.[18] Counterinsurgents will also need to be aware of the operations taking place in their area of responsibility, as well as elsewhere in the region, to understand the significance and objective of their present operations.

Although making the effort to learn language and culture is a valid activity in itself, it should also be seen as a pathway to unlocking the social nuances of the local population and may reveal potential cleavages to exploit within the insurgency. The insurgents are, after all, part of the population and who better to advise on these cleavages than the sea in which the insurgent swims? Once the counterinsurgents become reasonably acquainted with any cleavages within the insurgency, they should encourage competition among potential factions and may be able to convince individual insurgents to defect by granting marginal demands. In certain cases, such as the Philippines and Vietnam, counterinsurgent knowledge and exploitation of cleavages within the insurgency and the population as a whole abetted the pacification effort.

In the Philippines, Vietnam, and Colombia, the counterinsurgents were slow to understand the complexities of the insurgencies they faced, to the detriment of their COIN strategies. Because COIN is a system of programs designed to react and adapt to insurgency, it is important to understand the nature of the conflict as well as the strengths and weaknesses of the insurgents. Counterinsurgents at all levels of operation should seek to understand the insurgent cause so as to undermine it, or at least avoid giving it credibility through counterinsurgent actions. In this effort, the counterinsurgents should try not to cloud their perceptions with their prior belief systems or overreliance on analogies, as this may lead to a faulty perception of the nature and origins of the conflict, as well as insurgent grievances. Additionally, foreign counterinsurgents should support the wishes of the population above those of the host nation, so as to maintain the legitimacy of the counterinsurgent cause. If the wishes of the

[18] Although the insurgency in Iraq is not analyzed in this paper, pp. 23–29 of Ahmed Hashim's *Insurgency and Counterinsurgency in Iraq* discuss historical and sociological factors that caused Fallujah to become such a hotbed of instability in the insurgency.

population and the host nation are radically different, it will hurt the counterinsurgent cause to turn a blind eye to host nation corruption, nepotism, and inefficiency.

Forcing troops to live close to the population wherever possible can mitigate the perception of counterinsurgents as outsiders interested in their own security.[19] Because of their physical and sometimes psychological proximity to the population, the counterinsurgents in the Philippines, Vietnam (Marine Corps CAP units), and J&K were able to gather intelligence to target the insurgent infrastructure and were also able to limit the manpower and supply flows to the enemy. Such proximity to the population can help counterinsurgents better protect the civilian population and better understand what grievances need to be addressed to drain support from the insurgency.

Counterinsurgency training should be seen as a system and should be approached as such by the appropriate authorities. There should be an objective individual charged with powers of oversight to make sure not only that there is fusion and continuity among COIN programs, but that there are no gaps or redundancies as well. Kalev Sepp asserts that this person can either be a home-grown charismatic leader or a foreign adviser in the background who would ideally understand that COIN operations and conventional warfare require different skill sets and training.[20]

Counterinsurgents should seek to create a competent indigenous police force free from corruption, accompanied by reforms in the justice and penal systems. In recruiting indigenous security forces, the counterinsurgents should seek to create a force that reflects the ethnic, religious, and socioeconomic makeup of the local population and should make a special effort to recruit a reasonable number of potentially oppressed ethnic minorities to increase their stake in fighting the insurgency.[21] Indigenous police should be trained by other police forces who have been steeped in the appropriate institutional culture and capabilities rather than by the military. By no means should the military do police work, because the military is trained to capture and kill, whereas the police are trained to detain and interrogate. Police also protect the population from being coerced into joining the insurgency, as well as gather and disseminate intelligence gathered from interrogations. In the Philippines, the police force was instrumental in the capture of the insurgent leader, as well as in increasing the appearance of a gradual return of sovereignty to the Filipinos.

Foreign counterinsurgents should adopt a hands-off approach to operations—encouraging the host nation to develop its own capabilities and learn from its own successes and failures. This may sometimes require that counterinsurgents limit themselves to defensive operations, keeping in mind that if they offer too much assistance, the host nation will have few incentives to learn to develop its own procedures or change ineffective ones. As Cohen et al. (and earlier, T. E. Lawrence)[22] argue, it is better for the host nation to do a mediocre job than

---

[19] Tomes (2004, p. 24).

[20] Sepp (2005, p. 11).

[21] This applies to both the police and the military.

[22] "Do not try to do too much with your own hands. Better the Arabs do it tolerably than that you do it perfectly. It is their war, and you are to help them, not to win it for them. Actually, also, under the very odd conditions of Arabia, your practical work will not be as good as, perhaps, you think it is" (Lawrence, 1917).

for foreign counterinsurgents to do a perfect job in terms of the legitimacy that would be lost in the process.[23] Host nation visibility in the political, humanitarian, economic, and security spheres will help demonstrate its legitimacy and competence, with the foreign counterinsurgents playing only a supporting role.

Fledgling insurgencies are easiest to crush, as they are still formulating a cause, garnering support, and learning effective guerrilla techniques, and are thus more susceptible to failure.[24] Therefore, realizing the presence of a threat and defining what exactly that threat poses to the host nation and foreign counterinsurgents are critical procedures to undertake *before* the insurgency gains strength and momentum. In fact, crushing the insurgency while it is small can save time, money, and valuable lives, but to acknowledge a threat is to acknowledge the existence of a weakness within the current system that must be corrected.

The counterinsurgent loses if it does not win, which is why it is critical to develop programs to target the insurgent infrastructure early while it is still weak and vulnerable.[25] For this reason, counterinsurgents must have an uncorrupted view of the status of the insurgency and how to combat it at different stages in its development. When there is a lull in the insurgency, the insurgents may be lying low and have not necessarily been defeated. Therefore, counterinsurgents must be prepared to pursue insurgents until their ranks have been purged and so severely weakened that they can no longer pose a credible threat to the state apparatus. In the ongoing cases of J&K and Colombia, the counterinsurgents have been unable to completely eliminate the insurgent infrastructure and presence in the country. In Colombia, this has led analysts to assess that the FARC may not be in its dying days but may be resting and rebuilding for another round of fighting. In contemporary insurgencies, the task of eliminating insurgent infrastructure is perhaps more difficult because insurgent use of the Internet and other electronic modes of communication provide them with "e-sanctuary," which allows them to spread information and gain recruits in a highly protected medium.

To fight insurgencies, the counterinsurgents need to remember to become more primitive—not necessarily more sophisticated—in their techniques and to think outside the box. Advanced technology designed to minimize counterinsurgent casualties often prevents them from having to interact with the population and its abuse results in their alienation and radicalization. In the case of Vietnam, heavy use of indiscriminate firepower hurt the counterinsurgent effort rather than facilitating it, and the host nation in El Salvador relied too heavily on hand-me-down helicopters from the United States that often malfunctioned and required constant maintenance—a distraction when planning for and fighting the insurgency. In contrast, the counterinsurgents in Jammu and Kashmir have realized that they need to minimize the use of firepower and airpower to reduce civilian casualties. In any case, counterinsurgents should use technology as a way to gather additional information from the population

---

[23] Cohen (2006, p. 52).

[24] Byman (unpublished research).

[25] Cohen et al. (2006, p. 51).

during routine interactions, as well as to ensure the protection of the population from insurgent retribution.

Counterinsurgent adoption of preemptive detention and torture as extreme countermeasures to the insurgency in Algeria not only alienated moderates and drove them into the ranks of the insurgency but also caused a loss of support from the domestic population in France. Any political advances were undermined once people became aware of the atrocities committed by the counterinsurgents in pursuit of stability, and this ultimately forced France to grant Algeria independence. Similar human rights abuses in the form of death squads eroded government legitimacy in El Salvador, and U.S. soldiers in the Philippines undertook violent means, such as internment, food deprivation, and torture, to pacify the population. Preemptive and preventive detention can also radicalize previously unaffiliated civilians once they come in contact with insurgents who are being detained in the same prison.

To minimize insurgent use of sanctuary and the influx of manpower and supplies from other countries, counterinsurgents must make securing the border a high priority, depriving the insurgents of the strategic initiative and a place to recuperate and regroup. Counterinsurgent inability to deny the insurgents sanctuary proved to be a persistent problem in the cases of Vietnam, El Salvador, J&K, and Colombia. In Vietnam, the counterinsurgents resisted pursuing the enemy into bordering countries, fearing that such action would escalate the conflict. In J&K, the ambiguous areas of control and the decades-old tension between the nuclear nations of India and Pakistan make pursuing insurgents into their sanctuary in Pakistan suicidal. In the case of the insurgencies in Colombia, the insurgents not only receive sanctuary in neighboring Venezuela but are also receiving some level of support from Venezuelan President Hugo Chavez. On the other hand, the counterinsurgents were extremely successful in closing Algeria's borders to supplies and manpower, damaging the effectiveness of the remaining insurgents in Algeria.

The lessons learned from the cases in this paper exhibit several COIN TTPs that met with varying levels of success and failure. Factors of perception and mindset have not been discussed here but practitioners and theorists alike must keep in mind that fighting insurgency can be a prolonged affair, with few obvious successes or easy answers. To most effectively manage the complexity of COIN, counterinsurgents need to think of solutions in terms of long-term effectiveness, not short-term necessity—and this especially refers to the successes and failures from past COIN operations. Algeria is a prime example of a case in which the counterinsurgent perception of short-term solutions to immediately crush the insurgency counteracted its long-term goals of finding a political solution to the conflict.

Above all, although a multidisciplinary discussion drawing lessons from the past and present is certainly helpful in crafting a response to the threat of insurgency, creating a model for fighting insurgency can be destructive, as it may lead counterinsurgents to pigeon-hole their responses based on historical analogies that follow the model. This is not to say that we cannot learn from history. On the contrary, when put in perspective, having a base of knowledge of the outcomes of a mix of variables can be a valuable resource on which counterinsurgents can draw—allowing lessons learned to become money in the bank.

Bibliography

Andrade, Dale, and James H. Willbanks, "CORDS/Phoenix: Counterinsurgency Lessons from Vietnam for the Future," *Military Review*, March–April 2006, pp. 9–23.

Bailey, Cecil E., "OPATT: The U.S. Army SF Advisers in El Salvador," *Special Warfare*, December 2004. As of March 21, 2007:
http://findarticles.com/p/articles/mi_m0HZY/is_2_17/ai_n13824770

Blaufarb, Douglas S., *The Counterinsurgency Era: U.S. Doctrine and Performance 1950 to the Present*, New York: The Free Press, 1977.

Boot, Max, *The Savage Wars of Peace: Small Wars and the Rise of American Power*, New York: Basic Books, 2002.

Bracamonte, José Angel Moroni, and David E. Spencer, *Strategy and Tactics of the Salvadoran FMLN Guerrillas*, Westport, Conn., and London: Praeger, 1995.

Brennan, Richard, Adam Grissom, Sarah A. Daly, Peter Chalk, William Rosenau, Kalev Sepp, and Steve Dalzell, ongoing RAND Corporation research on future insurgency threats.

Byman, Daniel L., "Friends Like These: Counterinsurgency and the War on Terrorism," *International Security*, Vol. 31, No. 2, Fall 2006, pp. 79–115.

———, unpublished RAND research on understanding proto-insurgencies.

Carter, William Harding, *The Life of Lieutenant General Chaffee*, Chicago, Ill.: University of Chicago, 1917.

Cassidy, Robert M., "Back to the Street without Joy: Counterinsurgency Lessons from Vietnam and Other Small Wars," *Parameters*, Vol. 34, No. 2, Summer 2004, pp. 73–83.

Central Intelligence Agency, *Guide to the Analysis of Insurgency*, Washington, D.C.: Central Intelligence Agency, n.d.

Clausewitz, Carl von, *On War*, Baltimore, Md.: Penguin Books, 1968.

Coates, Maj Robert J., U.S. Marine Corps, "The United States' Approach to El Salvador," 1991. As of March 21, 2007:
http://www.globalsecurity.org/military/library/report/1991/CRJ.htm

Cohen, Eliot, et al. "Principles, Imperatives, and Paradoxes of Counterinsurgency," *Military Review*, March–April 2006, pp. 49–53.

"Colombia President Rides Popularity Wave: Uribe Widely Seen as Tough-Minded," *Los Angeles Times*, September 25, 2005.

Cooley, John, and Green March, *Black September: The Story of the Palestinian Arabs*, London: Frank Cass, 1973.

Council on Foreign Relations, "FARC, ELN, AUC (Colombia, Rebels)," November 2005. As of March 21, 2007:
http://www.cfr.org/publication/9272/

Deady, Timothy K., "Lessons from a Successful Counterinsurgency: The Philippines, 1899–1902," *Parameters*, Vol. 35, No. 1, Spring 2005, pp. 53–68.

Espejo, German, and Juan Carlos Garzon, "La Encrucijada del ELN," *Informe Especial,* Bogotá, Colombia: Fundacion Seguridad y Democracia, July 2005.

Frente Farabundo Martí para la Liberación Nacional, "Historia del FMLN," n.d. As of March 21, 2007:
http://fmln.org.sv/portal/index.php?module=htmlpages&func=display&pid=1

Fuerzas Armadas Revolucionarias de Colombia, "Las FARC: 30 Años de Lucha por la Paz, Democracia y Soberanía," May 27, 1994. As of May 24, 2007:
http://www.farcep.org/?node=2,1890,1

Fundación Seguridad y Democracia, "Balance de Seguridad 2005," 2006.

Galula, David, *Pacification in Algeria, 1956–1958*, Santa Monica, Calif.: RAND Corporation, MG-478-1-ARPA/RC, 2006. As of May 22, 2007:
http://www.rand.org/pubs/monographs/MG478-1/

Gaucher, Roland, *Les Terroristes*, Paris: Editions Albin Michel, 1965.

Gettleman, Marvin, et al., *El Salvador: Central America in the New Cold War,* New York: Grove Press, 1981.

Gompert, David, and John Gordon, unpublished RAND research on building complete and balanced capabilities for counterinsurgency.

Harding, William Carter, *The Life of Lieutenant General Chaffee*, Chicago, Ill.: University of Chicago, 1917.

Hart, Alan, *Arafat: A Political Biography*, London: Sidgwick and Jackson, 1994.

Hashim, Ahmed S., *Insurgency and Counterinsurgency in Iraq*, Ithaca, N.Y.: Cornell University Press, 2006.

Hirst, David, *The Gun and the Olive Branch*, London: Futura, 1977.

Hoffman, Bruce, *Inside Terrorism*, London: Victor Gallancz, 1998.

Horne, Alistair, *A Savage War of Peace, Algeria, 1954–1962*, London: Macmillan, 1977.

Hunt, Richard A., *Pacification: The American Struggle for Vietnam's Hearts and Minds*, Boulder, Colo.: Westview Press, 1995.

International Institute for Strategic Studies, *The Military Balance 2006*, 2006.

Joint Chiefs of Staff/Department of Defense, *Dictionary of Military and Associated Terms,* JP 1-02, April 12, 2001. As of May 14, 2007:
http://www.dtic.mil/doctrine/jel/new_pubs/jp1_02.pdf

Karnow, Stanley, *Vietnam: A History*, New York: Penguin Books, 1997.

Kilcullen, David, "Counterinsurgency Redux," *Survival*, Winter 2006–07, pp. 111–130.

Kruger, Alexander, "El Salvador's Marxist Revolution," The Heritage Foundation, Backgrounder No. 137, April 10, 1981. As of March 21, 2007:
http://www.heritage.org/Research/LainAmerica/mg137.cfm

Lawrence, T. E., "The 27 Articles," *The Arab Bulletin,* August 20, 1917. As of May 15, 2007:
http://net.lib/buy/edu/~rdh7/wwi/1917/27arts.html

Linn, Brian MacAllister, *The U.S. Army and Counterinsurgency in the Philippine War, 1899–1902*, Chapel Hill, N.C.: University of North Carolina Press, 1989.

———, *The Philippine War, 1899–1902*, Lawrence, Kan.:, University Press of Kansas, 2000.

Logan, Sam, "Guns, Cocaine: One Market Out of Control," *International Relations and Security Network (ISN) Security Watch,* Swiss Federal Institute of Technology (ETH Zurich), February 28, 2006. As of March 22, 2007:
http://www.isn.ethz.ch/news/sw/details.cfm?ID=14921

"Los Planes de las Farc," *Semana*, August 7, 2000.

Mandela, Nelson, *Long Walk to Freedom*, London: Abacus, 1994.

Maitre, H. Joachim, "The Dying War in El Salvador," in Walter F. Hahn and Jeane J. Kirkpatrick, eds., *Central America and the Reagan Doctrine,* Lanham, Md.: University Press of America, 1987.

Marighela, Carlos, "The Minimanual of the Urban Guerrilla," in Grant Wardlaw, ed., *Political Terrorism. Theory, Tactics and Counter-Measures,* Cambridge, UK: Cambridge University Press, 1982.

Martin, Gilles, "War in Algeria: The French Experience," *Military Review*, July–August 2005, pp. 51–57.

Maullin, Richard, *Soldiers, Guerrillas, and Politics in Colombia*, Boston, Mass.: D. C. Heath, 1973.

McKinley, William, letter to Wesley Merritt, May 19, 1898a. U.S. Army Adjutant General's Office, *Correspondence Relating to the War with Spain . . . April 15, 1898, to July 4, 1902*, 2 vols., Washington, D.C.: Center of Military History, reprint 1993.

———, letter to Secretary of War Russell A. Alger, Washington, D.C., December 21, 1898b.

McNamara, Robert S., James G. Blight, and Robert K. Brigham, *Argument Without End: In Search of Answers to the Vietnam Tragedy*, New York: Public Affairs, 1999.

Metz, Steven, and LTCOL Raymond A. Millen, *Insurgency and Counterinsurgency in the 21st Century: Reconceptualizing Threat and Response*, Carlisle, Pa.: Strategic Studies Institute of the U.S. Army War College, 2004.

Miller, Stuart Creighton, *Benevolent Assimilation: The American Conquest of the Philippines, 1899–1903*, New Haven, Conn.: Yale University Press, 1982.

Mukherjee, Anit, "Lessons from Another Insurgency," *New York Times*, March 4, 2006.

Nagl, John A., *Learning to Eat Soup with a Knife: Counterinsurgency Lessons from Malaya and Vietnam*, Westport, Conn.: Praeger Publishers, 2002.

National Security Archives, *El Salvador: The Making of US Policy, 1977–1984*, Alexandria, Va.: Chadwick-Healey, Inc., 1989.

O'Balance, Edgar, *Arab Guerrilla Power*, London: Faber, 1974.

Packer, George, "Knowing the Enemy: Can Social Scientists Redefine the 'War on Terror'"? *The New Yorker,* December 18, 2006.

Petraeus, LTGEN David H., "Learning Counterinsurgency: Observations from Soldiering in Iraq," *Military Review*, January–February 2006.

Pike, Douglas, *PAVN: People's Army of Vietnam*, Novato, Calif.: Presidio, 1986.

"Population Density by Country: Latin America and the Caribbean" (n.d.). As of March 22, 2007:
http://www.overpopulation.com/faq/basic_information/population_density/latin_america.html

Pribbenow, Merle L., *Victory in Vietnam: The Official History of the People's Army of Vietnam, 1954–1975*, Lawrence, Kan.: University Press of Kansas, 2002.

Rabasa, Angel, and Peter Chalk, *Colombian Labyrinth: The Synergy of Drugs and Insurgency and Its Implications for Regional Stability*, Santa Monica, Calif.: RAND Corporation, MR-1339-AF, 2001. As of March 22, 2007:
http://www.rand.org/pubs/monograph_reports/MR1339/

Rangel, Alfredo, *Colombia: Guerra en el Fin de Siglo*, Bogotá: Tercer Mundo, 1998.

Ray, Lieutenant General Arjun, *Kashmir Diary*, New Delhi: Manas Publications, 1997.

The Report of the President's National Bipartisan Commission on Central America, New York: Macmillan Publishing Company, 1984.

Robinson, Linda, "Terror Close to Home," *U.S. News and World Report*, Vol. 135, No. 11, October 6, 2003.

Rosie, George, *Directory of International Terrorism*, Edinburgh: Mainstream Publishing, 1986.

Roth, Russell, *Muddy Glory: America's "Indian Wars" in the Philippines 1899–1935*, Boston, Mass.: Christopher Publishing House, 1981.

Sardeshpande, Lieutenant General S. C., *War and Soldiering*, New Delhi: Lancer Publishers Pvt. Ltd., 1993.

Schiff, Zeev, and Raphael Rothstein, *Fedayeen: The Story of the Palestinian Guerrillas*, London: Valentine Mitchell, 1972.

Schulzinger, Robert D., *A Time for War: The United States and Vietnam, 1941–1975*, New York: Oxford University Press, 1997.

Sepp, Kalev I., "Best Practices in Counterinsurgency," *Military Review*, May–June 2005, pp. 9–12.

Sood, Lieutenant General V. K., and Pravin Sawhney, *Operation Parakram: The War Unfinished*, New Delhi: SAGE Publications, 2003.

South Asia Terrorism Portal, *Jammu and Kashmir Assessment*, 2003. As of April 2006:
http://www.satp.org

Suarez, Alfredo Rangel, "Balance Estratégico," *El Tiempo*, Colombia, September 2, 2005.

Thackrah, John Richard, *Encyclopedia of Terrorism and Political Violence*, London: Routledge Kegan and Paul 1987.

Tomes, Robert M., "Relearning Counterinsurgency Warfare," *Parameters*, Spring 2004, pp. 16–28.

Tovo, LTCOL Ken, *From the Ashes of the Phoenix: Lessons for Contemporary Counterinsurgency Operations*, Carlisle, Pa.: U.S. Army War College, 2005.

United States Institute of Peace, *Truth Commissions Digital Collection: Reports: El Salvador, Chronology of the Violence, 1993*. As of March 22, 2007:
http://www.usip.org/library/tc/doc/reports/el_salvador/tc_es_03151993_chron2_4.html

U.S. Department of the Army and U.S. Marine Corps, *Counterinsurgency*, FM 3-24/MCWP 3-33.5, Washington, D.C.: Department of the Army Headquarters, 2006.

U.S. Library of Congress, Federal Research Division, "Country Study: El Salvador—Background to the Insurgency," 1988a. As of March 22, 2007:
http://www.country-data.com/cgi-bin/query/r-4322.html

U.S. Library of Congress, Federal Research Division, "El Salvador: A Country Study—Foreign Military Influence and Assistance," November 1988b. As of March 22, 2007:
http://www.country-data.com/cgi-bin/query/r-4315.html

U.S. Library of Congress, Federal Research Division, "Country Study: El Salvador—Relations with the United States," n.d. As of March 26, 2007:
http://countrystudies.us/el-salvador/84.htm

Võ, Nguyên Giáp, *How We Won the War*, Philadelphia, Pa.: RECON Publication, 1976.

Wardlaw, Grant, *Political Terrorism, Theory, Tactics and Counter-Measures*, Cambridge, UK, and New York: Cambridge University Press, 1982.

Wikipedia Encyclopedia, "Algerian War of Independence," n.d. As of May 10, 2006:
http://en.wikipedia.org/wiki/Algerian_War_of_Independence

Wolff, Leon, *Little Brown Brother: How the United States Purchased and Pacified the Philippine Islands at the Century's Turn*, Garden City, N.Y.: Doubleday & Company, Inc., 1961.